Electrophysiology

Jürgen Rettinger · Silvia Schwarz ·
Wolfgang Schwarz

Electrophysiology

Basics, Methods, Modern Approaches and Applications

Second Edition

 Springer

Jürgen Rettinger
Tübingen, Baden-Württemberg
Germany

Silvia Schwarz
Shanghai University of Traditional Chinese Medicine
Shanghai Research Center for Acupuncture and
Meridians
Shanghai, China

Wolfgang Schwarz
Institute for Biophysics
Goethe University
Frankfurt, Hessen
Germany

ISBN 978-3-030-86484-2   ISBN 978-3-030-86482-8   (eBook)
https://doi.org/10.1007/978-3-030-86482-8

This Springer imprint is published by the registered company Springer Nature Switzerland AG
The registered company address is: Gewerbestrasse 11, 6330 Cham, Switzerland

Preface

This book is based on the previous versions by J. Rettinger, S. Schwarz, and W. Schwarz entitled *Electrophysiology* and *Elektrophysiologie,* which form the basis for lectures on Electrophysiology held at Goethe-University Frankfurt, Chinese Academy of Sciences (Shanghai), and Fudan University Shanghai. This new second edition of the English version is again addressed to students of biology, chemistry, and physics with special interest in biophysics. Because of the heterogeneity of the addressed readers, we will try to give basic information on physical as well as biological background but not more than necessary.

Tübingen, Baden-Württemberg, Germany   Jürgen Rettinger
Shanghai, China   Silvia Schwarz
Frankfurt am Main, Hessen, Germany   Wolfgang Schwarz
June 2021

The original version of this book was revised: Electronic Supplementary video files are included in Chapter 1.

Acknowledgements

We would like to express our thanks to Prof. Guanghong Ding and Prof. Dr. Di Zhang from Fudan University Shanghai and to Prof. Dr. Werner Mäntele and Dr. Georg Wille from Goethe University Frankfurt for their continuous support in having Electrophysiology a part of education in Biophysics at the two universities. This was and still is the basis for this book.

About This Book

After a brief introductory and historical overview on electrophysiology (see Chap. 1), basic electrochemical principles for the understanding of this subject are summarised (see Chap. 2). Thereafter, electrophysiological methods including principles of data analysis are presented in Chap. 3 ranging from measurements on the whole animal via measurements on single cells with microelectrodes to the patch-clamp technique. This chapter also includes a brief presentation of ion-selective microelectrodes, the carbon-fibre technique, and the sniffer-patch method. The basic method in electrophysiology is the voltage-clamp technique (see Chap. 4), and different ways of applying the voltage clamp are described. The chapter deals with different versions of the conventional two-microelectrode technique, the patch-clamp technique, and examples for automated electrophysiology. In Chap. 5, the major conductance pathways in cell membranes, the ion-selective channels, are described with respect to their specific characteristics. Essentials of membrane excitability based on Hodgkin-Huxley description of an action potential and synaptic transmission are presented in Chap. 6. The sensitivity of modern voltage-clamp techniques allows detection of carrier-mediated transport, and Chap. 7 presents by means of three examples characteristics of carriers compared to channels and how electrophysiological methods can be used for functional characterisation. Finally, in Chap. 8 we finish with exemplary illustrations of how combination of electrophysiology, molecular biology, and pharmacology can be applied to learn about the structure, function, and regulation of membrane permeabilities that form the basis of cellular function and how they are involved in diseases. In addition to the Na,K pump and the GABA transporter, as examples for active transporters, the purinergic receptor P2X and viral ion channels are introduced as examples for ion channels. A special section illustrates how electrophysiology can be used to understand basic cellular mechanisms in Chinese medicine and to investigate drug-receptor interaction in pharmacology. Each chapter is completed by a "Take-Home Messages" table and a set of exercises for recalling important topics.

The book is supplemented by appendices (Chap. 9) describing the influence of electrical and magnetic fields on physiological function and a manual for a Laboratory Course in electrophysiology using the two-electrode voltage clamp.

Important Physical Units

In the following we will list important electrical quantities and their usual definitions.

Voltage U [Volt, V]:   1 V is the difference in electric potential between two points of a conductor carrying a constant current of 1 A, when the power dissipated between those points is equal to 1 W.

Resistance R [Ohm, Ω]:   1 Ω is the resistance of a conductor such that a constant current of 1 A produces a voltage drop of 1 V between its ends.

Conductance g [Siemens, S]:   The inverse of the resistance *R*.

Current I [Ampere, A]:   1 A is that constant current which, if maintained in two straight parallel conductors of infinite length, of negligible circular cross section, and placed one metre apart in a vacuum, would produce between these conductors a force equal to 2×10^{-7} Newton per 1 m of length.

Charge Q [Coulomb, C]:   1 Coulomb equals the charge of 6.25×10^{18} elementary charges *e*.

Capacitance C [Farad, F]:   1 F is the capacitance of a capacitor between the plates of which there appears a difference of potential of 1 volt when it is charged by a quantity of charge equal to 1 coulomb.

Magnetic Flux Density B [Tesla, T]:   1 T is the density of a homogenous magnetic flux of 1 Weber (Wb) perpendicular to an area of 1 m²: $1\text{T} = 1 \text{ Wb/m}^2$.

Volt: $V = \frac{W}{A} = \frac{\text{kg}\cdot\text{m}^2}{\text{A}\cdot\text{s}^3}$

Ohm: $\Omega = \frac{V}{A} = \frac{\text{kg}\cdot\text{m}^2}{\text{A}^2\text{s}^3}$

Siemens: $S = \frac{1}{\Omega} = \frac{\text{A}^2\text{s}^3}{\text{kg}\cdot\text{m}^2}$

Current: $A = A$
Coulomb: $C = A \cdot s$
Farad: $F = \frac{C}{V} = \frac{A^2 s^4}{kg \cdot m^2}$

The quantities listed above can be also expressed in terms of SI units (metre, kilogram, second, Ampere).

Note: Only charge and current are expressed in terms of SI units (A and s).

Contents

About the Authors

Jürgen Rettinger received his doctor degree in physics at Goethe University in Frankfurt in 1995. His focus as group leader at Max-Planck-Institute of Biophysics in Frankfurt was research of structure-function relationships of ligand-gated ion channels and teaching electrophysiology at Goethe-University Frankfurt, Shanghai Institutes for Biological Sciences, and Fudan University Shanghai. Between 2008 and 2021 Jürgen Rettinger worked with Multi Channel Systems responsible as product manager for automated electrophysiology.

Silvia Schwarz finished her university education in biology and chemistry at Albert-Ludwigs-University Freiburg in 1974. Thereafter, she was educated and worked as teacher in biology and chemistry until 2008. Since then she works as research associate at Shanghai Research Center for Acupuncture and Meridians (Shanghai Key Laboratory of Acupuncture Mechanism and Acupoint Function, Fudan Univ.). Silvia Schwarz's main research focus is on antiviral effects of Chinese herbal drugs applying electrophysiological methods.

Wolfgang Schwarz received his doctor degree in physics at the University of Saarland in 1975. Thereafter, he spent two years as postdoc at the University of Washington Seattle working on characteristics of ion channels. In 1981 he habilitated in physiology at the University of Saarland and received a position as head of a Cell Physiology Laboratory at Max-Planck-Institute of Biophysics, Frankfurt. Wolfgang Schwarz's main focus turned to structure-function and regulation of carrier transporters and teaching electrophysiology as professor for physics at Goethe-University Frankfurt, Shanghai Institutes for Biological Sciences, and Fudan University Shanghai. Since 2004 Wolfgang Schwarz is head of the Cell Electrophysiology Laboratory of Shanghai Research Center for Acupuncture and Meridian (Shanghai Key Laboratory of Acupuncture Mechanism and Acupoint Function, Fudan Univ.) and concurrent professor at Fudan University.

Abbreviations

γ	Single-channel conductance, activity coefficient
δ	Quantisation step, charge density
ε_0	Polarisability of free space
η	Viscosity
λ	Length constant
μ	(Electro)chemical potential
ρ	Specific resistance
τ	Time constant
a	Activity
A	Gain (op-amplifier), area
ASM	Airway smooth muscle
AU	Arbitrary unit
au	Arbitrary units
B	Bandwidth
C	Capacitance
c	Concentration
Ca^{2+}_i	Intracellular calcium activity
CypA	Cyclophilin A
D	Diffusion coefficient, dissipation factor (capacitor)
DOR	δ-opioid receptor
DPDPE	[D-Pen^2,D-Pen^5]enkephalin
e	Elementary charge
E, V, Φ	Electric potential
EAAC	Excitatory amino acid carrier
EC	(−)-epicatechin
EGCG	(−)-epigallocatechin-3-gallate
F	Faraday's constant
f	Frequency
FBA	Feedback amplifier
G	Gibb's energy
GA	Glycyrrhizic acid

GABA	γ-amino buteric acid
GAT	GABA transporter type
g	conductance
h	Planck's constant
HH	Hodgkin-Huxley
I	Current
i	Single-channel current
J	Flux, current density
k	Boltzmann's constant, rate constant
l, a, x	Length, distance, depth
N_a	Avogadro's constant
NXC	Na^+, Ca^{2+} exchanger
p	Probability, dipole momentum, permeability
Q	Charge
r	Resistivity
R	Universal gas constant
R, R_Ω	Resistance
S	Spectral density
T	Temperature
t	Time
TCM	Traditional Chinese Medicine
TEVC	Two-electrode voltage clamp
TMS	Transmembrane segment
TRPV	Transient-receptor-potential channel sensitive to vanilloid
U	Energy
v	Velocity
z	Valency

Introduction

1

Contents

Abstract

This chapter shall give an overview and some basic background on what will be presented in the following chapters. The introduction will also give a brief overview on the history of electrophysiology. Particularly the modern developments shall then be dealt with in more detail in the following chapters.

This and each of the following chapters are completed by *Take-Home Messages* and a set of *Exercises* for recalling important topics.

Keywords

Lipid bilayer · Carrier proteins · Channel proteins · History of electrophysiology

The first part of the introductory chapter will give an overview and some basic background on what we shall present in the following chapters. The introduction shall also give a brief overview on the history of electrophysiology. In particular, the modern developments shall be dealt with in more detail in the later chapters. For detailed information on the

Supplementary Information The online version contains supplementary material available at [https://doi.org/10.1007/978-3-030-86482-8_1].

The original version of this chapter was revised: Electronic Supplementary Video files are included at https://doi.org/10.1007/978-3-030-86482-8_10

electrophysiology of ion channels we would like to direct the reader to the excellent text books by Hille (2001) and Aidley and Stanfield (1996). We also like to refer to the *Axon Guide* (Sherman-Gold, 2008). For illustrating practical application of Two-Electrode Voltage Clamp, we prepared a supplementary video (Videos 1–4) using the *Xenopus* oocytes as a cell model system.

1.1   Basic Background Knowledge

Electrophysiology is a very powerful biophysical method dealing with the investigation of electrical properties of cell membranes and their functional significance. The different functional roles cells play in an organism are determined to a high degree by specific proteins in or at the cell membrane. We shall not describe the structure and components of a cell membrane in detail, but we briefly want to remind the reader about basic characteristics.

The cell membrane is composed of a phospholipid bilayer, which has a high electrical resistivity of about 10^{15} Ωcm (see Table 2.2, Sect. 2.1) and separates cytoplasm from extracellular space. Cell-specific proteins are embedded in or attached to the membrane (Fig. 1.1).

These proteins are highly specific in their function and serve for communication between extracellular space and cytoplasm, along the cell membrane, or between cells. The structure of the membrane is not rigid, but the bilayer rather represents a fluid-crystalline mosaic (Singer & Nicolson, 1972). Proteins become localised by cytoskeletal structures. The interaction of the proteins with cytoskeleton or other cytoplasmic components plays a significant role in regulation of function of the proteins.

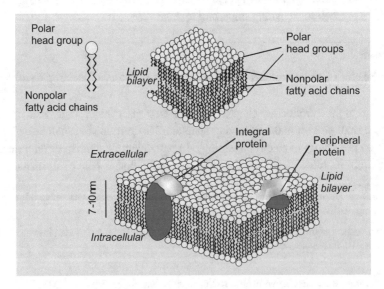

Fig. 1.1   Structure of a cell membrane as a fluid-crystalline phospholipid bilayer (according to Singer & Nicolson, 1972) with embedded proteins

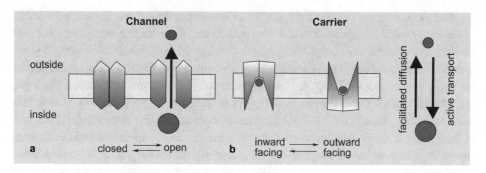

Fig. 1.2 Schematic illustration of channels (**a**) and carriers (**b**). The size of the red balls represents the electrochemical potential of the respective substrate

Table 1.1 Characteristic properties of ion channels and carriers

	Pores	Carriers		
Type of transport	Ion channels	Facilitated diffusion	Primary active transport	Secondary active transport
Driving force	Diffusion along a gradient	Diffusion along a gradient	Transport against a gradient utilising	
			Metabolic energy (ATP)	Ion gradients
Rate of transport	$<10^8$ s^{-1}		$<10^3$ s^{-1}	
Conductance	~1–300 pS		$\ll$pS	
Function	Signalling function		Housekeeping function	

For proper function of the cell nutrients, metabolites and ions have to be transported across the hydrophobic cell membrane. This is achieved by special groups of membrane proteins, the transport proteins. These transport proteins can be classified into two major categories (see Fig. 1.2), such as channels (or pores, Chap. 5) and carriers (Chap. 7).

The classical channel is a protein that can form a pore, and in its open state it allows ions to diffuse along their electrochemical gradient across the membrane. The transition between an open and a closed pore involves conformational changes within the protein and is called gating. Once a pore is open, ions can cross the membrane at rates of up to about 10^7–10^8 s^{-1}. This rate corresponds to single-channel currents of up to ~20 pA at the resting potential and to conductances of up to 300 pS (see Sect. 4.4.3). For a carrier, each translocation of a substrate involves a series of conformational changes. Since such changes usually occur in the range of milliseconds or even slower, translocation rates of less than 10^3 s^{-1} are typical for carriers leading to conductances in the sub-fS range. The average rate of transport for both systems can, nevertheless, be similar since channels are usually open only for very brief time periods in the ms range while carriers operate continuously. Channels, therefore, are often involved in fast signalling mechanisms while carriers serve for housekeeping (see Table 1.1).

If transport proteins mediate net transport of ions across the cell membrane, we will call them *electrogenic* transporters. In this view ion channels are electrogenic per se, but also carriers can transfer net-charge. The current flowing through channels or generated by carriers can be involved in cellular signalling; the current may directly act as signal (like in an action potential (see Chap. 6)) or may indirectly affect cellular function (through, e.g. modulation of signalling pathways). The lipid bilayer represents a resistor with a very low conductance (10^{-8} S/cm^2) between cytoplasm and the extracellular fluid, and consequently a small membrane current can generate an electrical potential difference across the membrane in the range of several tens of mV (this is simplified called *membrane potential*). Changes in membrane conductance (predominantly based on the opening or closing of ion channels) will lead to changes in membrane potential. These changes in membrane potential are essential for many cellular functions, like the spread of action potentials in excitable cells. In addition, they may directly affect the membrane currents generated by electrogenic transporters (where the potential represents part of the driving force) or also indirectly by influencing the conformational changes in the protein that are involved in regulation of the transport.

In addition to electrogenic transporters, carriers exist that do not transport a net-charge either that an electrically neutral molecule is transported or that compensating charge is co- or counter-transported. Such transporters, nevertheless, can be regulated by membrane potential (see e.g. Sect. 7.1.3). If the conformational changes necessary for substrate translocation involve charge movements within the protein or if ion binding occurs within the electrical field, then membrane potential will affect transport activity provided these steps are rate-determining.

It should be pointed out that "rate-determining" does not necessarily mean "rate-limiting".

Electrophysiology nowadays deals with different techniques of analysing electrical signals (current and potential changes) generated at cell membranes or with techniques where electrical stimuli are guided to the membrane to analyse their influence on membrane transport and cell function. Electrical signals coupled to cell membranes can often be detected at the body surface of an animal or can be applied to the body surface. It is also possible to investigate isolated tissue, single cells or even isolated membrane patches. These techniques will be described in later sections (Chap. 3).

1.2   History of Electrophysiology

This chapter is based on the excellent review written by C.H. Wu (Wu, 1984). The first bioelectrical phenomena mankind was confronted with were the discharges from the electrical organ of certain fishes. Already in the ancient Egypt the cat fish from Nile River was known, which can generate voltage pulses of up to 350 V.

From the Mediterranean Sea five species of the electric ray are known; in Fig. 1.3, the electric ray *Torpedo torpedo* can be identified on a mosaic from Pompeii of the first century. The *Torpedo* can generate with its electric organ voltage pulses of 45 V.

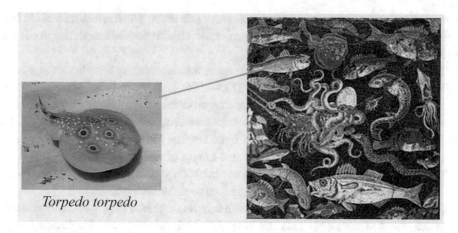

Fig. 1.3 Mosaic from Pompeii (first century p.Chr.) showing *Torpedo torpedo* (Museo Archeologico Nazionale, Naples, https://www.flickr.com/photos/carolemage/14820098532/)

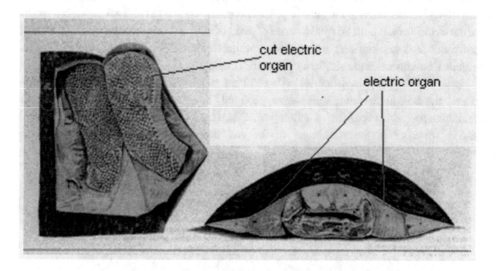

Fig. 1.4 Structure of electric organ of *Torpedo* (based on Hunter, 1773–1774)

At these ancient times nothing was known about the basis of those phenomena, however, the discharges generated by these animals were utilised in a way, which we would nowadays call *electrotherapy*. Detailed instructions on how to use the *Torpedo* were written down by Scribonius Largus (see Sconocchia, 1983) who lived at the time of Emperor Claudius (41–54 p.Chr.). Scribonius Largus, e.g. described the treatment of headache or gout by placing the *Torpedo* on the body surface or the extremities into a water container with the *Torpedo*. Over the centuries these therapeutic instructions were distributed all over Europe.

First attempts to understand the phenomena were developed in mid of the 18th century when physicians started to investigate the anatomy of the electric organ (Fig. 1.4). The organ is composed of columns, and each column of thin disks. When scientists realised that these disks originated from muscle cells, very controversial speculations started about the electrical nature of the discharging signals.

The demonstration of an electrical phenomenon was reported by John Walsh in a letter to Benjamin Franklin (Walsh & Seignette, 1773–1774). Later on in 1776 John Walsh even succeeded to make the discharge of the electric organ visible in the form of a light flash, and finally convinced even the most critical scientists about the electrical nature of the discharge phenomena; we may place the birth hour of electrophysiology to this year. End of the 18th century, Alessandro Volta and Luigi Galvani demonstrated, though from different points of view, that electrical phenomena not only form the basis of the function of the electric organ but also in general of all the nerve and muscle activity.

Galvani (1791) first showed that frog muscle contracts when the muscle or the innervating nerve was touched with a bi-metal arch (Fig. 1.5). Galvani interpreted his observation in analogy to the electric organ as a discharge of electrical energy stored in the muscle cells.

In contrast, Volta, confronted with Galvani's idea in 1792, believed that the use of two metals produced an electrical potential difference between the two ends of the arch, and the electricity is transferred to the muscle cells. This interpretation led him to construct a model of the electric organ built up from a series of disks of zinc and copper separated by pieces of sea water soaked cloth and arranged in a vertical column (Volta, 1800, see Fig. 1.6a). Figure 1.6b illustrates the similarity between Volta's pile and the electric organ.

Later it was recognised that all cells exhibit a potential difference across their membrane, the so-called resting membrane potential, and that in nerve and muscle cells excitation spreads in the form of action potentials, very brief changes in polarity that travel like a wave along the membrane of the cell fibre. Basis of these phenomena are

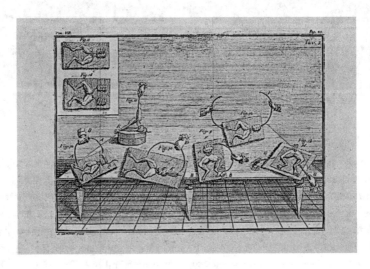

Fig. 1.5 Experiment by Galvani illustrating excitability of nerve-muscle preparation (from Galvani, 1791)

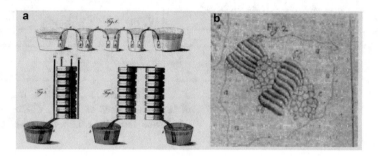

Fig. 1.6 Electric pile described by Allessandro Volta in 1800 (**a**), columns of electrocytes (**b**, based on Lorenzini, 1678)

electrochemical gradients for ions that can pass the cell membrane. These ions are predominantly Na^+, K^+, and Cl^-, and in a resting cell the sum of currents, I, carried by these ions is zero (steady state):

$$I_{Na} + I_K + I_{Cl} = I = 0.$$

The ion fluxes are driven by the respective concentration and electrical gradients, and can be described by the Nernst–Planck equation (see Sect. 2.4). With the assumptions of independent ion movements with a constant diffusion coefficient, and of a constant electrical field in the membrane, the Nernst–Planck equation can be integrated leading to the often-applied Goldman–Hodgkin–Katz (GHK) equation for the zero-current (=resting) membrane potential E:

$$E = \frac{RT}{F} \ln \left(\frac{P_{Na}[Na^+]_o + P_K[K^+]_o + P_{Cl}[Cl^-]_i}{P_{Na}[Na^+]_i + P_K[K^+]_i + P_{Cl}[Cl^-]_o} \right).$$

We should emphasize that here and also in later equations not the term concentration should be used, but to be acurately the term activity that is equal to the concentration of dissociated ions according to:

$$activity = \gamma * concentration$$

with activity coefficient γ. Because activity has also the dimension of "concentration" and often ratios of activities are of relevance, we often will use for simplicity the term concentration that is usually also used in physiology. But we should keep in mind that in physiological solution γ is well below 1 (see Table 1.2).

In terms of the GHK equation, the potential difference at a cell membrane is governed by the ion-specific permeabilities (see Sect. 2.3). $P = Da$ with diffusion coefficient D and membrane thickness a.

Ion species	100 mM	10 mM	1 mM
Na^+	0.78	0.902	0.969
K^+	0.76	0.899	0.964
Ca^{2+}	0.40	0.675	0.87
Cl^-	0.76	0.899	0.964

Table 1.2 Activity coefficient γ in solutions of different ionic strength[a]

[a]Data from "Activity Coefficients at 25 °C. (2020). Retrieved March 27, 2021, from https://chem.libretexts.org/@go/page/44398

After it was discovered that excitability of nerve and muscle cells is based on the spread of action potentials along the cell fibres, Bernstein (Bernstein, 1902, 1912) hypothesised that the resting potential is determined by a selective permeability for K^+ ions, and the action potential originates due to a breakdown of ion selectivity. In fact, as it turned out later, during an action potential the membrane potential becomes even positive inside compared to outside.

A qualitatively new step was introduced to electrophysiology by Allan Hodgkin and Andrew Huxley (Hodgkin & Huxley, 1952). The two British scientists succeeded to demonstrate that the phenomena of electrical excitability and the generation of an action potential can be attributed to specific alterations of ion conductances (Chap. 6), a result that was honoured by the Nobel Prize in 1963. Basis for this work was the so-called voltage-clamp technique first introduced by Cole (1949) and Marmont (1949), and further developed by Hodgkin and Huxley together with Katz (Hodgkin et al., 1949, 1952).

An essential step in Hodgkin's and Huxley's work was the separation of the ion-specific current components, which allowed them to investigate the time- and voltage-dependent conductances for Na^+ and K^+ ions that form the basis of the time course of an action potential. The Hodgkin–Huxley description (see Sect. 6.1.2) led to the question of how conductances become voltage-dependent. A possible interpretation was the hypothesis of ion-selective channels that can exist in an open or in a closed state, where the transitions between these states involve charge movements within the channel protein. Another question deals with the conductance of the pore in its open state. Rough estimation assuming that an ion can freely diffuse with its hydration shell across the membrane through a water filled pore suggested that a channel should exhibit a conductance of less than 300 pS (see Sect. 4.4.3), which would mean that currents of the order of 10 pA will flow through a single open channel. It was impossible in those days to detect such tiny currents with the conventional voltage-clamp technique since unspecific leak currents as well as current noise were already by orders of magnitude larger. Another milestone in electrophysiology, therefore, was the development of a new voltage-clamp technique, the *patch-clamp method* by the two German scientists Erwin Neher and Bert Sakmann (Neher & Sakmann, 1976, Nobel Prize in 1991) that allowed them to detect currents through single channels in isolated membrane patches (see Sect. 4.4).

The highlights we mentioned above and further important discoveries that led to nowadays electrophysiology are summarised in Table 1.3.

Table 1.3 Highlights related to electrophysiological research

Time	Name	Subject
Before 2750 a. Chr.	*Reliefs in Egyptian tombs*	*First indications of electric phenomena*
44–48 p.Chr.	*Scribonius Largus*	*Application of electric organ of Torpedo to medical treatment (in Scribonii Largi de compositione medicamentorum liber)*
1773	*J. Hunter*	*Morphology of the electric organ*
1776	*J.Walsh*	*Demonstration of electric nature by an electric spark (starting point of electrophysiology)*
1791	*L. Galvani and A. Volta*	*Nerve and muscle excitation*
1906	**C. Golgi and S. Ramon y Cajal**	**Work on the structure of the nervous system**
~1910	*J. Bernstein*	*Hypothesis for action potential*
1920	**W. Nernst**	**Work in thermochemistry**
1924	**W. Einthoven**	**Discovery of the mechanism of the electrocardiogram**
1932	**E.D. Adrian C. Sherrington**	**Discoveries regarding the functions of neurons**
1936	**H.H. Dale and O. Loewi**	**Discoveries relating to chemical transmission of nerve impulses**
1944	**J. Erlanger and H.S. Gasser**	**Discoveries relating to the highly differentiated functions of single nerve fibres**
1949	**W.R. Hess**	*Discovery of the functional organization of the interbrain as a coordinator of the activities of the internal organs*
1949	*Cole and Marmont*	*Development of voltage-clamp technique*
1961	**G. v. Békésy**	**Discoveries of the physical mechanism of stimulation within the cochlea**
1963	**J. Eccles, A.L. Hodgkin and A.F. Huxley**	**Discoveries concerning the ionic mechanisms involved in excitation and inhibition in the peripheral and central portions of the nerve cell membrane**
1967	**R. Granit, H.K. Harline and G. Wald**	**Discoveries concerning the primary physiological and chemical visual processes in the eye**
1970	**J. Axelrod, B. Katz and U. v. Euler**	**Discoveries concerning the humoral transmitters in the nerve terminals and the mechanism for their storage, release and inactivation**
1981	**R.W. Sperry and D.H. Hubel, T.N.Wiesel**	*Discoveries concerning the functional specialization of the cerebral hemispheres and information processing in the visual system.*
1991	**E. Neher and B. Sakmann**	**Discoveries concerning the function of single ion channels in cells**
1994	**A.G. Gilman, M. Rodbell**	**Discovery of G-proteins and the role of these proteins in signal transduction in cells**

(continued)

Table 1.3 (continued)

Time	Name	Subject
1997	P.D. Boyer, J.E. Walker and J.C. Skou	Elucidation of the enzymatic mechanism underlying the synthesis of adenosine triphosphate (ATP) and discovery of an ion-transporting enzyme, Na^+, K^+ - ATPase
2000	A. Carlson, P. Greengard, E.R. Kandel	Discoveries concerning signal transduction in the nervous system
2003	P. Agre and R. MacKinnon	Discovery concerning channels in cell membranes (discovery of water channels and structural and mechanistic studies of ion channels)
2004	R. Axel, L.B. Buck	Discoveries of odorant receptors and the organization of the olfactory system
2021	D. Julius, A. Patapoutian	Discoveries of receptors for temperature and touch

Work that was honoured by Nobel Prize is printed in bold

Take-Home Messages

1. The experience with **electrical phenomena** is dating back to the time of ancient Egypt (**nearly 5000 years ago**).
2. As **birth hour of electrophysiology**, we can consider the demonstration of **electric basis** by Walsh of what we now call the "electric organ" **end of eighteenth century**.
3. The application of **voltage-clamp technique** by Hodgkin and Huxley (**in the 1950s**) to describe an action potential by **voltage- and time-dependent ionspecific permeabilities** was the mile stone for modern cell electrophysiology.
4. Another key technique in electrophysiology, demonstrating single-channel events, is the **patch-clamp technique** and was developed in the **early 1970s** by Neher and Sakmann.
5. The **activity coefficient** γ (activity = γ concentration) **differs under physiological conditions significantly from 1**.

Exercises

1. Recall dates of milestones in history of electrophysiology, and the respective discoveries or statements.
2. Discuss the question: Who was right, Luigi Galvani or Alessandro Volta?
3. How can animals generate high voltage pulses? Describe the electrophysiological basis?

References

Aidley, D. J., & Stanfield, P. R. (1996). *Ion channels, molecules in action*. Cambridge University Press.

Bernstein, J. (1902). Untersuchungen zur Thermodynamik der bioelektrischen Ströme. *Pflügers Archiv, 92*, 521–562.

Bernstein, J. (1912). *Elektrobiologie*. Vieweg.

Cole, K. S. (1949). Dynamic electrical characteristics of the squid axon membrane. *Arch Sci Physiol, 3*, 253–258.

Galvani, L. (1791). De viribus electricitatis in motu musculari commentarius. *Bononiensi Scientiarum et Artium Instituto atque Academia Commentarii, 7*, 363–418.

Hille, B. (2001). *Ionic channels of excitable membranes* (3rd ed.). Sinauer Associates Inc..

Hodgkin, A. L., & Huxley, A. F. (1952). A quantitative description of membrane current and its application to conductance and excitation in nerve. *The Journal of Physiology, 117*, 500–544.

Hodgkin, A. L., Huxley, A. F., & Katz, B. (1949). Ionic currents underlying activity in the giant axon of the squid. *Archives Science Physiology, 3*, 129–150.

Hodgkin, A. L., Huxley, A. F., & Katz, B. (1952). Measurements of current-voltage relations in the membrane of the giant axon of Loligo. *The Journal of Physiology London, 116*, 424–448.

Hunter, J. (1773–1774). Anatomical observations on the Torpedo. *Philosophical Transactions, 63*, 481–489.

Lorenzini, S. (1678). *Osservazioni intorno alle torpedini fatte*. Per l'Onofri.

Marmont, G. (1949). Studies on the axon membrane I. A new method. *Journal of Cellular and Comparative Physiology, 34*, 351–382.

Museo Archeologico Nazionale, Neapel, https://www.flickr.com/photos/carolemage/14820098532/

Neher, E., & Sakmann, B. (1976). Single-channel currents recorded from membrane of denervated frog muscle fibres. *Nature, 260*, 799–802.

Sconocchia, S. (1983). *Scribinii Largi compositions*. Teubner.

Sherman-Gold. (2008). *The axon guide for electrophysiology and biophysics laboratory techniques* (3rd ed.). Axon Instruments.

Singer, S. J., & Nicolson, G. L. (1972). The fluid mosaic model of the structure of cell membranes. *Science, 175*, 720–731.

Volta, A. (1800). On the electricity excited by the mere contact of conducting substances of different kinds. *Philosophical Transactions (in French), 90*, 403–431.

Walsh, J., & Seignette, S. (1773–1774). On the electric property of the torpedo. *Philosophical Transactions, 63*, 461–480.

Wu, C. H. (1984). Electric fish and the discovery of animal electricity. *American Scientist, 72*, 598–607.

Basics Theory

2

Contents

Abstract

Basic electrochemical principles essential for the understanding of electrophysiology shall be presented in this chapter. This topic includes electrical characteristics of biological membranes and the distribution of ions inside and outside a cell. The activity gradients for different ion species form the basis for all electrical phenomena at a cell membrane. Different ion-selective permeabilities will lead to membrane potential that can be described by various equations.

Keywords

Ion distribution · Nernst equation · Donnan potential · Goldman–Hodgkin–Katz equation

2.1 Electrical Characteristics of Biological Membranes

Parameters often used in describing electrophysiological phenomena are listed in Table 2.1. These physical constants will be used throughout.

© The Author(s), under exclusive license to Springer Nature Switzerland AG 2022
J. Rettinger et al., *Electrophysiology*,
https://doi.org/10.1007/978-3-030-86482-8_2

Table 2.1 Often used physical constants

Constant	Abbreviation	Value	Unit
Avogadro's constant	N_A	$6.022 \cdot 10^{23}$	mol^{-1}
Elementary charge	e	$1.602 \cdot 10^{-19}$	C
Boltzmann's constant	k	$1.381 \cdot 10^{-23}$	$J\,K^{-1}$
Universal gas constant	$R\ (kN_A)$	8.314	$J\,K^{-1}\,mol^{-1}$
Faraday's constant	$F\ (eN_A)$	$9.648 \cdot 10^4$	$C\,mol^{-1}$

Table 2.2 Typical values of resistivity

		Extracellular solution		
	Lipid bilayer	for mammals	for amphibians	Sea water
$r\ (\Omega cm)$	10^{15}	60	80	20

The distribution of particles of potential energy U is governed by the Boltzmann law:

$$c(U) = c_0 e^{-U/kT}.$$

Therefore, the Boltzmann factor $e^{U/kT}$ ($= e^{\Delta EzF/RT}$ with electrical potential gradient ΔE and effective valency z), which describes charge distribution in an electrical field, plays an important role in electrophysiological descriptions. It is, therefore, very useful to know the value of RT/F in mV. At room temperature (293 °K) the value is

$$\mathbf{RT/F \approx 25\ mV} \quad \mathbf{or} \quad \mathbf{\ln{(10)}\ RT/F \approx 58\ mV}.$$

In electrophysiology distribution between two orientations is often considered, and the charge distribution Q is then described by the Fermi equation:

$$Q(\Delta U) = \frac{1}{1 + \exp{(\Delta EzF/RT)}}.$$

Other basic physical rules and characteristics are:

1. *Ohm's Law*: $E = I \cdot R_\Omega$ or $I = g \cdot E$; I denotes current, R_Ω resistance, and the conductance g is $1/R_\Omega$.
2. *Resistivity* (r), which is a measure for the electric resistance. It is defined by $R_\Omega = r\,l/F$ with length (e.g. of nerve fibres) l and sectional area F. r is usually given in Ωcm. Typical values are given in Table 2.2 illustrating again the high electrical resistivity of the lipid bilayer compared to electrolyte solution at physiological ion activities.
3. *Capacitance* ($C = Q/E$): Interestingly, the specific capacitance of a cell membrane does hardly depend on the cell type and is close to the capacitance of a pure lipid bilayer

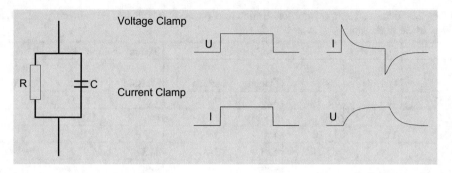

Fig. 2.1 Simplified electrical diagram of a membrane symbolised by a resistor R in parallel with a capacitor C, and responses to voltage(U)- or current (I)-clamp pulses, respectively

(ca. $0.8 \, \mu F/cm^2$). A value of $1 \, \mu F/cm^2$ is often used to estimate the surface area of a cell from electrical determinations of its capacitance. The capacitance can be obtained from the transient signal of charging or discharging the membrane capacitor.

A simplified model of a membrane (Fig. 2.1) consists of the parallel arrangement of a capacitor and a resistor. The capacitor C represents the capacitance of the lipid bilayer and the resistor R combines all conductance pathways across the cell membrane.

A transient signal in response to a voltage- or current-clamp pulse can be measured, and analysed in terms of (compare Fig. 2.1) :

$$I_C = \frac{dQ}{dt} = C\frac{dE}{dt} \Rightarrow \frac{dE}{dt} = \frac{I_C}{C} \Rightarrow dE = \frac{1}{C}I_C dt.$$

The charging/discharging of the capacitor follows an exponential time course:

$$E = E_0 e^{-t/\tau} \text{ with time constant } \tau = R\,C.$$

Typical values for a biological membrane with specific membrane resistances of 10–$10^6 \, \Omega cm^2$ are accordingly in the range of $10 \, \mu s$ to $1 \, s$.

2.2   Ion Distribution at Biological Membranes

All electrical phenomena occurring at a cell membrane are based on asymmetrical ion distributions between cytoplasm and extracellular space, and on ion-selective permeabilities or conductances. Table 2.3 gives an overview for the concentrations of the most relevant inorganic ions outside and inside of cells of three different animal species, which have been used extensively in electrophysiological investigations.

Table 2.3 Intra- and extracellular ion distributions (in mM) of cells that were investigated extensively in electrophysiology

	Squid axon		Frog muscle		Mammalian muscle		Ratio
							Outside
	Outside	Inside	Outside	Inside	Outside	Inside	Inside
Na^+	460	50	120	9.2	145	12	~10
K^+	10	400	2.5	140	4	155	~40^{-1}
Ca^{2+}	11	$3 \cdot 10^{-4}$	1.8	$3 \cdot 10^{-4}$	1.5	$<10^{-4}$	~10^4
Cl^-	540	40–100	120	3–4	123	4.2	*Variable*
E_R/mV	−60		−90		−90		

The table also gives rough estimates for measured resting membrane potentials (by convention: inside–outside potential). The distribution of the ions at cells of different origins shows qualitative similarities. In the cytoplasm we find about ten times less Na^+ compared to outside, but extracellularly we have more than one order of magnitude lower K^+. For Ca^{2+} there is also an inwardly directed gradient, and the activities differ by even four orders of magnitude. The extremely low intracellular activity of Ca^{2+} in sub-μM range can easily be increased temporarily either by influx of extracellular Ca^{2+} or by emptying intracellular Ca^{2+} stores. Such mechanisms are used by nature to regulate a large variety of cellular functions that depend on intracellular Ca^{2+} (Ca^{2+}_i).

The dominating extracellular anion is Cl^-. Intracellularly, the negative counter charges are predominantly negative charges on proteins and account for bulk electroneutrality. Electroneutrality is a basic principle that governs all electrophysiological processes. The total activities of ions inside compared to those outside the cell are identical.

The ion gradients are of physiologically essential importance. The question for the basis of the asymmetry in ion species gradient and its maintenance and for the functional consequences is of particular interest. In the following section we will consider first the basic electrochemical consequences. For further details see e.g. Hille (2001).

2.3   Donnan Distribution and Nernst Equation

Like all thermodynamic systems, the composition of the cytoplasm of a cell separated from its surrounding by the membrane approaches steady state, and thermal forces are in equilibrium with other forces. For chemical reactions and transport processes across the cell membrane this means that forward and backward reactions are identical if there is no additional energy input.

2.3.1   Donnan Distribution

Let us consider two compartments (O and I, referring to outside and inside, respectively) separated by a K^+- and Cl^--permeable membrane with rigid walls (see Fig. 2.2). To one of the water-filled compartments we add KCl. After some time, there will be equal distribution of the ion activities in the two compartments of, let us say, 100 mM. Via two Ag/AgCl electrodes (see Sect. 3.4.1) the electrical potential difference can be measured. Now we add to compartment I 50 mM of a salt KA, where the anion A^- cannot penetrate the membrane. In analogy to the condition in a living cell, this could represent the non-permeant protein anions in the cytoplasm.

The activity of K^+ in I is now higher than in O so that K^+ will diffuse along its gradient from I to O. A basic rule of electrophysiology is the *principle of electroneutrality*; as consequence an anion will have to follow the K^+, which will be Cl^- as the only permeant anion. In steady state, K^+ in I will still be higher than in O, but Cl^- will be lower. This distribution is named Donnan equilibrium (Donnan, 1911). In steady state, influx ($O \rightarrow I$) of KCl will equal efflux ($I \rightarrow O$). Since the rate of influx is proportional to $[K_O]\,[Cl_O]$, and the rate of efflux to $[K_I]\,[Cl_I]$, we have at steady state:

$$[K_I]\,[Cl_I] = [K_O]\,[Cl_O].$$

For the activities used in the above example we obtain (with x representing the amount of anions or cations that diffused from I to O or vice versa) for the steady-state condition:

$$(150 \ - x)\,(100 \ - x) = (100 + x)\,(100 + x)$$

$$\Rightarrow x \approx 11.$$

Due to the non-permeant anions, an electrical potential difference inside negative will develop at the membrane like at a capacitor, and this potential difference is called Donnan

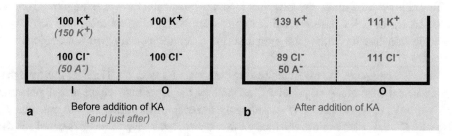

Fig. 2.2 Illustration of an experiment with a membrane permeable for K^+ and Cl^- (dashed line) separating the compartments I and O. The numbers may represent ion activities in mM. (**a**) represents situation before and (**b**) after addition of the K^+ salt KA

Potential (E_d) with side **I** negative compared to **O**. The activities of the permeant ions on side **I** can then be described by Boltzmann distribution:

$$[K_I] = [K_O] \ \exp\left(-E_dF/RT\right)$$

$$[Cl_I] = [Cl_O] \ \exp\left(+E_dF/RT\right)$$

(compare with above steady-state condition: $[K_O] [Cl_O] = [K_I] [Cl_I]$).
For E_d we obtain

$$E_d = -(RT/F) \ \ln([K_I]/[K_O]).$$

Since for side **I** the condition

$$[K_I] = [Cl_I] + [A^-]$$

holds, and for electroneutrality

$$[K_O] = [Cl_O]$$

has to be fulfilled, we obtain with $[K_O] [Cl_O] = [K_I] [Cl_I]$:

$$[K_I] \ ([K_I] - [A^-]) = [K_O] \ [Cl_O] = [K_O]^2$$

$$[K_I] = \tfrac{1}{2} [A^-] \pm \left([K_O]^2 + \tfrac{1}{4} [A^-]^2\right)^{1/2}$$

$$E_d = -(RT/F) \ \ln([K_I]/[K_O]) = -(RT/F) \ \ln\left[\frac{[A]}{2[K_o]} + \left(\left(\frac{[A]}{2[K_o]}\right)^2 + 1\right)^{1/2}\right].$$

This allows estimating the possible Donnan potentials for the examples given in Table 2.3 on the basis of known concentrations for the impermeant anions A^- (estimated from electroneutrality within the cytoplasm) and for the equally permeant anions and cations.

Table 2.4 compares the measured resting potentials E_m with E_d. The data illustrate that the measured membrane potential E_m as well as the amplitude of an action potential (ca. +30 mV, compare Fig. 2.4) considerably differ from the value for the potential E_d given by Donnan distribution; obviously other phenomena have to be considered for the generation of a membrane potential at rest and during excitation.

Table 2.4 Measured membrane potentials and those calculated from the data in Table 2.3

(mV)	Squid axon	Frog muscle	Mammalian muscle
E_m	−60	−90	−90
E_d	−9	−14	−16
E_K	−92	−100	−91
E_{Na}	+55	+64	+62
E_{Cl}	−51	−85	−84

With E_m (resting membrane potential, E_d (Donnan potential), E_K, E_{Na}, E_{Cl} (respective Nernst potentials)

2.3.2 Nernst Equation

Let us replace the membrane of our model system in Fig. 2.2 by a membrane that is permeable only for one ion species. The ratio of probabilities for finding a permeant ion on side **I** or side **O** (Fig. 2.2) can be described by Boltzmann distribution (Sect. 2.1):

$$p_I/p_O = \exp\left(-z(E_i - E_O)F/kT\right);$$

accordingly, we obtain for the ratio of ion activities on both sides

$$c_I/c_O = \exp\left(-z(E_i - E_O)F/RT\right).$$

The potential difference given by the activity gradient in steady state is the so-called Nernst potential ΔE_N (Nernst, 1888a, b):

$$\Delta E_N = -RT/zF \, \ln\left(c_I/c_O\right).$$

The Nernst potentials for the gradient of the most relevant ions are also listed in Table 2.4. Comparison with the measured resting potentials E_m yields that E_m is far away from E_{Na} but has a value between E_K and E_{Cl}. This suggests that biological membranes at rest have ion permeabilities that are different for the different ion species being more permeable for K^+ and Cl^- than for Na^+.

The principal difference between Donnan and Nernst potential is that a *Donnan potential* will be established if *all ions* are *permeable except for one ion species*, and a *Nernst potential* if *all ions* are *impermeable except for one ion species*.

2.4 Goldman–Hodgkin–Katz Equation

Small ions are usually highly hydrophilic particles that cannot cross lipid bilayer. Transport across the cell membrane is mediated by transport proteins, which are embedded in the membrane but have hydrophilic domains the ions can interact with. These domains can be considered as sites where the ions can temporarily bind to while crossing the membrane. Mathematical description of membrane transport can be approached in two ways:

(a) *Discrete description*: This strategy assumes several separate binding sites, and to cross the membrane the ions have to hop from binding site to binding site. This hopping can be described by the theory of absolute reaction rates (Glasstone et al., 1941; see Sect. 5. 1.2).

(b) *Continuous (classical) description*: This strategy is based on the concept of free diffusion assuming that the ions pass a homogenous membrane (Goldman, 1943).

In the introduction we have already pointed out that the net current I at steady state has to vanish for passively diffusing ions:

$$I_{Na} + I_K + I_{Cl} = I = 0$$

if Na^+, K^+, and Cl^- were the only permeant ions. The current carried by a single ion species of activity c is driven by two forces: the activity gradient dc/dx and the electrical potential gradient dE/dx. The electrochemical potential is defined by

$$\frac{d\mu}{dx} = \frac{d}{dx}[\mu^o + RTln(c)] + zF\frac{dE}{dx}$$

with μ^o the chemical potential at standard conditions. This can also be written as:

$$\frac{d\mu}{dx} = RT\frac{dc}{dx}\frac{1}{c} + zF\frac{dE}{dx}.$$

Accordingly, the current I_c for each ion species c can be described by the Nernst–Planck equation (Nernst, 1888a, b; Planck, 1890a, b) provided those ions move independently:

$$I_c = \frac{zFD_x}{RT}c\frac{d\mu}{dx} \text{ or } I_c = -zFD_x\left[\frac{dc_x}{dx} + \left(\frac{zFc_x}{RT}\right)\frac{dE}{dx}\right]$$

with ion-specific diffusion coefficient D, which is associated with mobility u according to the Nernst–Einstein relationship

$$D = \frac{RT}{F}u.$$

The flux components driven by the chemical gradient Φ_c and by the electrical gradient Φ_E are

$$\Phi_C = -D\frac{dc}{dx} \text{ and } \Phi_E = -\text{zuc}\frac{dE}{dx}.$$

In case of only one ion species and because of $I = 0$, integration leads to the known Nernst equation (see Sect. 2.3) for a diffusion potential:

$$E = E_{\text{Nernst}} = \frac{RT}{zF} \ln\left(\frac{c_o}{c_i}\right) = -\frac{RT}{zF} \ln\left(\frac{c_i}{c_o}\right)$$

considering $\frac{dc}{dx} = c\frac{d\ln(c)}{dx}$.

In case of several permeant ion species, the Nernst–Planck equation has to be integrated for each ion species separately, which can also be written in the form:

$$I_c = -\frac{zFD_x}{e^{zEF/RT}}\frac{d\left(c_x e^{zEF/RT}\right)}{dx} = -zF\frac{c_i e^{-zEF/RT} - c_o}{\int_0^a \frac{e^{zEF/RT}}{D_x} dx}$$

with the boundary conditions $E(0) = E_o = 0$ and $E(a) = E_i = -E$ on the outside and inside of the membrane, respectively.

Three assumptions are necessary to integrate the Nernst–Planck equation:

1. the already made assumption of *independent* ion movements,
2. a constant diffusion coefficient (*homogenous membrane phase*), and
3. that potential E varies linearly in the membrane of thickness a (*constant field*).

This leads to the Goldman–Hodgkin–Katz (GHK) equation (Goldman, 1943; Hodgkin & Katz, 1949) for the current, occasionally also called "constant field equation":

$$I_c = (zF)^2 \frac{E}{RT}\frac{D_c}{a}\frac{\left[c_i e^{zEF/RT} - c_o\right]}{e^{zEF/RT} - 1}.$$

According to independent ion movement the current I_c can be described as the sum of unidirectional fluxes $-I_{\text{in}}$ and I_{eff}:

$$I_c = -I_{\text{in}} + I_{\text{eff}}$$

with $I_{\text{in}} \propto c_o$ and $I_{\text{eff}} \propto c_i$. Hence, we obtain

$$I_{\text{eff}} = \frac{(zF)^2 E}{RT} P_c \frac{c_i}{1 - e^{-zEF/RT}}$$

$$I_{\text{in}} = \frac{(zF)^2 E}{RT} P_c \frac{c_o}{1 - e^{+zEF/RT}}$$

with ion-specific permeability coefficient $P_c = D_c/a$.

For the ratio of efflux to influx, we obtain the so-called Ussing flux ratio (Ussing, 1949):

$$\left| \frac{I_{\text{eff}}}{I_{\text{in}}} \right| = \frac{c_i}{c_o} e^{zEF/RT}.$$

This flux ratio has been tested in several preparations successfully. But in several others, deviations from Ussing equation were found, which shows that the above assumptions are not fulfilled and discrete ion movement may provide a better description (see Sect. 5.1.2).

An important procedure in electrophysiology is the analysis of current–voltage dependencies (IV relationships), which allows, e.g. characterisation of channel gating, mechanisms of transport or drug action. While Ohm's Law predicts a linear relationship ($I=V/R_\Omega$ or $I = E/R_\Omega$), the GHK equation predicts linearity only for symmetric concentrations ($c_i = c_o = c_{\text{sym}}$) with

$$I = \frac{(zF)^2}{RT} P \cdot c_{\text{sym}} \cdot E \quad \text{and} \quad R_\Omega = \frac{RT}{(zF)^2} \frac{1}{P \cdot c_{\text{sym}}}.$$

In all other cases nonlinear dependency is obtained (see Fig. 2.3) and can be described by variable slope or chord conductance.

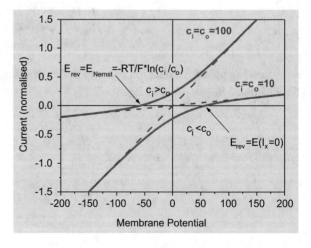

Fig. 2.3 Current–Voltage relationships for one ion species according to the GHK equation. c_i and c_o represent the intra- and extracellular ion activities, respectively

To determine the current–voltage dependence for a particular ion species, the respective current component has to be extracted from total membrane current. This can be achieved by blocking all other conductance pathways or by blocking the respective current and determining the difference of membrane current in the absence and presence of the inhibitor. Of course, also these procedures are based on the independence principle. Curved IV relationships are in fact often observed experimentally. But we should keep in mind that the derivation of the GHK equation is based on three assumptions (see above) that are all questionable.

For a single permeant ion species, a reversal of current direction is observed at $I_{in} = I_{eff}$ or $I_c = 0$, and the reversal potential is given by the Nernst equation for this ion species. If several conductance pathways contribute to the membrane potential, the reversal potential can be calculated by summing up the expressions of the GHK equation for the respective currents and setting the sum equal to 0. In our simple case for Na^+-, K^+-, and Cl^--selective pathways one obtains the GHK equation for the potential:

$$E_{GHK} = E_{rev} = \frac{RT}{F} \ln \left(\frac{P_{Na}[Na]_o + P_K[K]_o + P_{Cl}[Cl]_i}{P_{Na}[Na]_i + P_K[K]_i + P_{Cl}[Cl]_o} \right).$$

For a single ion species, this expression immediately takes the form of the Nernst equation. The GHK equation for the potential allows to calculate the membrane potential if relative permeabilities and concentrations are known or to determine relative permeabilities from measurements of the membrane potential at given ion concentrations. In the following we will give two simple examples for application of the GHK equation:

1. *The action potential*: According to Hodgkin and Huxley (1952), the permeabilities for Na^+ and K^+ of excitable membranes are time- and potential-dependent (see Sect. 6.1.2). At rest the membrane is predominantly permeable for K^+, and the membrane potential will be close to the Nernst potential for K^+, but slightly more positive according to

$$E_{GHK} = E_{rev} = \frac{RT}{F} \ln \left(\frac{P_{Na}[Na]_o + P_K[K]_o}{P_{Na}[Na]_i + P_K[K]_i} \right)$$

with $P_K \gg P_{Na}$. During development of the action potential, the Na^+ permeability increases dramatically leading to $P_{Na} \gg P_K$, and at the maximum of excitation the membrane potential approaches, therefore, the Nernst potential for Na^+ (see Fig. 2.4). With spontaneous inactivation of Na^+ permeability and gradual increase of K^+

Fig. 2.4 Simple explanation of
action potential according to
GHK equation for potential with
time- and potential-dependent
permeabilities for Na$^+$ and K$^+$.
E_{rest} (resting membrane
potential), E_K and E_{Na}
the respective Nernst potentials

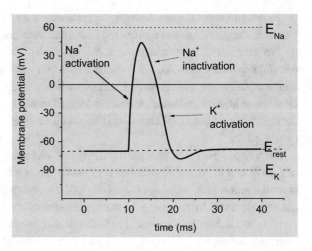

permeability, the membrane potential returns to its resting value. The entire process
occurs in the range of milliseconds.

2. *Bi-ionic conditions* for the determination of permeability ratios: If we have only one
   permeant ion species A on the outside and only one other permeant ion species B on the
   inside (bi-ionic condition), the reversal potential can be described according to GHK
   equation by

$$E_{rev} = \frac{RT}{F} \ln \left(\frac{P_A[A]_o}{P_B[B]_i} \right).$$

In the simple case of $[A]_o = [B]_i$, the reversal potential directly gives the permeability
ratio between A and B. Differences in permeability may originate from different
ion-specific pathways or from a common pathway being not equally permeable for
the two ion species. This example already shows how questionable the independence
principle can be under certain circumstances.

Take-Home Messages
1. Important biophysical values to remember are:

$$RT/F = 25 \text{ mV } C_{\text{specific}} \approx 1 \text{ }\mu F/cm^2.$$

2. **Lipid bilayer** membranes are almost **impermeable to ions.**
3. Essential for cell life is the **activity gradient** of various ions across the cell membrane (in the **millimolar range**), e.g.

$$Na^+_{\text{out}}/Na^+_{\text{in}} \approx 10.$$
$$K^+_{\text{out}}/K^+_{\text{in}} \approx 40^{-1}.$$
$$Ca^{2+}_{\text{out}} \gg Ca^{2+}_{\text{in}} \left(< 10^{-6}M \right).$$

4. All electrogenic, transmembrane processes obey the **Principle of Electroneutrality**, e.g. $[K^+_o] [Cl^-_o] = [K^+_i] [Cl^-_i]$.
5. **Nernst potential** will be established if a membrane is **permeable for only one ion species.**

   Donnan potential will be established if a membrane is **impermeable for one ion species.**
6. The **GHK equations** are based on **three assumptions**
   (a) Independence of ion movements.
   (b) Constant diffusion coefficient.
   (c) Linear potential change within the membrane (constant field).

   Though all the assumptions are **not realistic**, the equation can be used with restriction for qualitative descriptions (e.g. time course of action potential, permeability ratios).
7. A **simple electric description** of a membrane is **a resistor parallel to a capacitor**, representing the conductance pathways and the lipid bilayer, respectively.

Exercises

1. What is the value of RT/F at room temperature in mV, and why is it an important biophysical parameter?
2. What is the specific membrane capacitance, and why is it an important biophysical parameter?

3. What are the ratios of intra- to extracellular ion activities for a selected cell type? List the typical ion activities for mammalian and amphibian cells and for cells from marine species.
4. Under which conditions can we expect a Donnan potential, under which ones a Nernst potential at a cell membrane?
5. Calculate the Donnan potential for the ionic conditions listed in Table 2.3.
6. What are the assumptions the GHK equation is based on, and how is it derived?
7. Write down the GHK equation for the current. What are the characteristics of a current–voltage dependency based on GHK equation?
8. Derive the GHK equation for the potential from the current equation. Discuss the permeability changes during an action potential.

References

Donnan, F. G. (1911). Theorie der Membrangleichgewichte und Membranpotentiale bei Vorhandensein von nicht dialysierenden Elektrolyten. *Zeitschrift für Elektrochemie, 17*, 572–581.

Glasstone, S. K., Laidler, J., & Eyring, H. (1941). *The theory of rate processes*. McGraw-Hill Book Company.

Goldman, D. (1943). Potential, impedance and rectification in membranes. *The Journal of General Physiology, 27*, 37–60.

Hille, B. (2001). *Ionic channels of excitable membranes* (3rd ed.). Sinauer Associates.

Hodgkin, A. L., & Katz, B. (1949). The effect of sodium ions on the electrical activity of the giant axon of the squid. *Journal of Physiology (London), 108*, 37–77.

Hodgkin, A. L., & Huxley, A. F. (1952). A quantitative description of membrane current and its application to conductance and excitation in nerve. *The Journal of Physiology, 117*, 500–544.

Nernst, W. (1888a). Die elektromotorische Wirksamkeit der Ionen. *Zeitschrift für Physikalische Chemie*, 129–181.

Nernst, W. (1888b). Zur Kinetik der in Lösung befindlichen Körper. *Zeitschrift für Physikalische Chemie*, 613–637.

Planck, M. (1890a). Ueber die Erregung von Elektricität und Wärme in Elektrolyten. *Annals of Physical Chemistry, 39*, 161–186.

Planck, M. (1890b). Ueber die Potentialdifferenz zwischen zwei verdünnten Lösungen binärer Elektrolyte. *Annals of Physical Chemistry, 40*, 561–576.

Ussing, H. H. (1949). The distinction by means of tracers between active transport and diffusion. Transfer of iodide across isolated frog skin. *Acta Physiologica Scandinavica, 19*, 43–56.

Basics: Methods

3

Contents

Abstract

Electrophysiology deals with the analysis of electrical properties and signals, which can be studied in biological preparations. The essential electrophysiological methods including the principles of data analysis shall be presented in this chapter. The techniques cover measurements on the whole animal as well as measurements on single cells with microelectrodes down to the patch-clamp technique. This chapter also includes brief presentation of ion-selective microelectrodes, the carbon-fibre technique, and the sniffer-patch method.

© The Author(s), under exclusive license to Springer Nature Switzerland AG 2022   27
J. Rettinger et al., *Electrophysiology*,
https://doi.org/10.1007/978-3-030-86482-8_3

Keywords

Electrocardiogram · Ussing chamber · Brain slice · Field potential · Ag/AgCl electrode · Ion-selective electrode · Carbon-fibre electrode · Voltage-clamp technique · Sniffer patch

Electrophysiology deals with the analysis of electrical properties and signals which can be studied in biological preparations. In the previous chapter we have presented some theoretical background that is essential for describing electrical phenomena at the cell membrane. In this chapter we want to talk about basic methods that will allow to gain an understanding of the electrophysiological phenomena. When we discussed the GHK equation for the potential (Sect. 2.4), we briefly mentioned how in excitable cells (nerve and muscle cells) the action potential is governed by changes in permeability ratios. Also in other cells the membrane potential is determined by voltage- and time-dependent permeabilities, and these mechanisms essentially determine the function of a cell. Modifications in their properties often indicate dysfunction involved in diseases or they have physiologically regulatory function.

For analysis of permeability changes at a cell membrane, different techniques are available for the scientist. We want to introduce and classify now techniques by starting with methods that can be used in a living animal (for predominately medical-diagnostical (e.g. electrocardiogram (ECG)) or therapeutical (e.g. electroshock) purposes) going down to methods that allow investigation of the function of single membrane proteins, which are nowadays used extensively in elucidating structure–function relationships as well as dysfunction.

3.1 Recording Electrical Signals from Body Surface

Different techniques for recording electrical signals from the body surface have been established in medical diagnosis. Some of the techniques are listed in Table 3.1. Recording from body surface is possible since the animal body, simply speaking, represents an electrolyte (150 mM NaCl) container. With electrodes attached to the body surface tiny electrical signals can be detected that originate from electrical activity inside the container and that are conducted to the surface. At the beginning of the last century the Dutch physiologist Willem Einthoven succeeded to detect with a string galvanometer and a projection microscope (Einthoven, 1925, Nobel Prize, 1924) characteristic electrical potential oscillations that correlated with heart rhythm.

During the following decades it could be shown that these signals indeed originate from the activity of the beating heart. Therefore, such recordings were named electrocardiogram or briefly ECG.

Figure 3.1 illustrates such typical potential changes that can be detected by recording between left leg and the two arms and that repeat with the heart rhythm. Different recording

Table 3.1 Electromedical techniques

For diagnosis		For therapy	
Electrodermatography		Cardiac Resynchronization Therapy (CRT)	
Electrogastrography (from surface or intragastric)		Electroshock (defibrillation)	
Electrocochleography		Electrosleeptherapy	
Electrocardiography (from surface or intracardial)	*ECG*	Electrophysiology (Electroacupuncture)	
Electroencephalography	*EEG*	Electrogymnastics	
Electroneurography (conduction pattern)	*ENG*	Electrocoagulation	
Electrospinography		(Electronarcosis) (Electrotetanustherapy)	*No longer used because of uncontrolled damage*
Electrooculography Electroretinography Electronystagmography	*EOG*		
Electromyography	*EMG*		

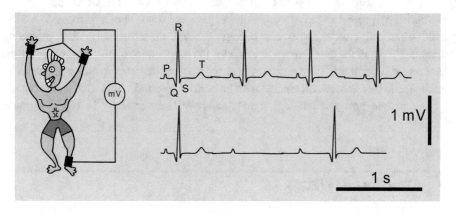

Fig. 3.1 Symbolised potential changes (ECG) recorded from a human that repeat with heartbeat, which show the normal PQRST waves (upper trace) and waves in the presence of atrioventricular block (lower trace)

sites are nowadays used, and international standardisation was introduced like the location of the recording sites, speed of recording device, time constants, and filters of the amplifier system.

Before we will deal with the ECG in more detail, we briefly want to remind the reader by Fig. 3.2 of the function of the heart as a pump.

The venous blood returning from the body enters the right atrium and is then transported via the right ventricle to the lung were CO_2/O_2 exchanges occurs. The O_2-loaded blood enters the left atrium and is then pumped by the strong left ventricle through the large body circuit.

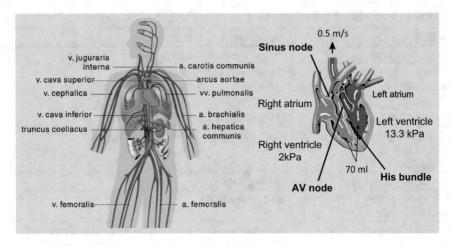

Fig. 3.2 Function of the heart (based on CorelDraw ClipArt)

The contraction of a heart muscle fibre, like in skeletal muscle, is induced by an action potential that spreads over the membrane of the cell fibre leading to an elevation of cytoplasmic Ca^{2+} activity. Important for proper function contraction of the heart muscle is controlled by proper spread of excitation in space and time. This is different to skeletal muscle where all muscle fibres can become excited simultaneously. Fast excitation-conducting fibres in the heart are responsible for the proper control of spread of excitation. Therefore, at a particular time a certain area is in an excited state while other areas are unexcited, and this pattern will change with time. With electrodes placed at particular sites (e.g. at the right arm and left leg) signals as shown schematically in Fig. 3.1 can be recorded.

3.2   The Example (ECG)

3.2.1   Electrophysiological Basics

We want to discuss now the basis of how these signals originate (for details see e.g. Katz, 1977). Let us consider the membrane of a muscle fibre. Due to the ion-selective membrane permeabilities and different concentrations of the ions in extracellular space and cytoplasm, the resting membrane has an inside negative potential compared to the outside or an outside positive potential compared to the inside of about 80 mV. At a site of excitation an action potential will be generated. According to our previous discussion, the permeability ratios will change in such a way that the membrane potential becomes temporarily depolarised, and the outside will even become negative compared to the inside (about −50 mV, see Fig. 3.3).

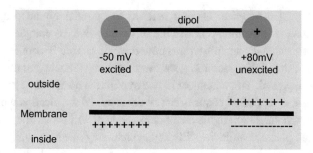

Fig. 3.3 Symbolised momentary state of spread of excitation at a cell membrane that can be interpreted from the outside as an electrical dipole changing with time. Note that the polarity commonly used for medical recording from body surface is opposite (outside compared to inside) to normal electrophysiological nomenclature (inside compared to outside)

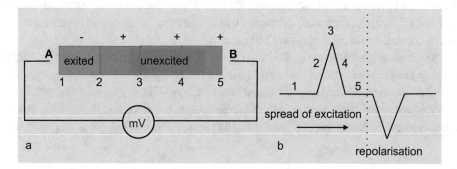

Fig. 3.4 The muscle fibre as an electrical dipole (**a**) and the change in potential difference (**b**) between the two ends of the fibre while excitation spreads from A to B. Five different phases of excitation are illustrated and labelled by 1–5

Looking from the outside, the excited area is negative compared to the resting area. Positive charges are opposing negative charges. With this qualitatively acceptable simplification we can treat this section of a cell membrane like an electrical dipole. In a similar way, we want to look at an entire muscle fibre in a momentary state of excitation. Let us consider first an unexcited simplified muscle fibre (comp. Fig. 3.4a, b).

Two electrodes **A** and **B** are to be located extracellularly. The surface will be positive everywhere compared to the cytoplasm, and there will be no potential difference between **A** and **B**. Now an excitation is induced at the position of **A**, leading to an action potential travelling with time into direction of **B**. After a short time period, a certain section of the muscle fibre will be excited, and the outside will now be negative. The cell looks like a dipole, and we will record a potential difference between **A** and **B**. As a general rule, the site where excitation is initiated is considered to be the reference so that the potential difference will be positive. The further the excitation spreads along the cell fibre the larger will be the potential difference. Having reached the middle of the cell, the potential difference will have a maximum and decline thereafter. For the completely excited cell, the potential difference will again be zero; now the cell surface is continuously negative. With this kind of extracellular recording, we cannot distinguish whether the entire cell is at rest or completely excited. This statement is also true for the heart when recording an ECG; only time- and space-dependent changes can be detected. At the location where the muscle was excited first, repolarisation of the membrane potential will occur first. Here the outside will become again positive while the still excited areas stay outside negative. We will now record a change to negative potential difference with a minimum when half of the cell has reached its resting state, and again zero when the entire cell is unexcited. Now the cycle can start again.

In the heart not only a single cell becomes activated, but also we have pointed out already, excitation spreads in a characteristic way over the entire heart. Again, excited (that means negative) areas will oppose resting (that means positive) areas. The distribution of charges can be described at least qualitatively by a simple dipole that will show changes in time and orientation. In general, electrodes will not be brought into direct contact with the heart but rather can be placed on the body surface.

The situation is schematically illustrated in Fig. 3.5. A dipole (**p**) in the centre represents a certain state of excitation of the heart in the body. The right half in Fig. 3.5 is unexcited and hence positive, the left half excited and hence negative. The dipole is located within our body, which consists to a large extent of electrolyte solution, and simplified, can be considered as a homogenous conductor (electrolyte container). In Fig. 3.5 the equipotential

Fig. 3.5 The symbolised heart as a dipole in the centre of an electrolyte container with equipotential lines (dotted) (left). The magnitude of dipole moment in direction of detection (forming an angle α with the dipole) is given by the projection on direction of detection: $p \cos\alpha$ (right)

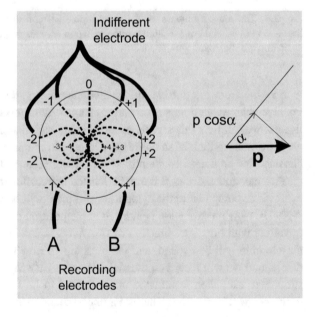

lines are shown as dotted lines. Each point on a given line has the same electrical potential. The lines in the half of the negative pole indicate negative potentials and *vice versa*.

The dependence of the potential (*E*) on distance (*r*) from the dipole with momentum $p = qa$ (where *a* is the distance vector between the charges *q*) can be described in a very simplified manner assuming (1) an infinite homogenous medium and (2) that $a \gg r$ by

$$E = \frac{1}{4\pi\varepsilon\varepsilon_0} \frac{p\cos\alpha}{r^2}$$

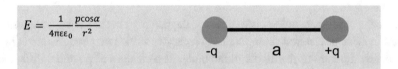

with α representing the angle between *a* with the direction of detection. These are obviously assumptions that are not true for the heart in our body for a quantitative description, however, sufficient for a qualitative description. The potential between two surface points is accordingly represented by the projection of the dipole vector on the direction of recording. If the recording direction is perpendicular to *p*, there will be no potential difference.

3.2.2   Activation of the Heart Muscle

During the excitation process in the heart, the dipole vector will vary with time in magnitude and orientation, and this synchronously with the heart cycle. Let us try to describe the movement of such a vector during the excitation of the ventricles.

Initiated by specialised excitation-conducting cells, activation will start from the septum of the left ventricle initiated by the activity of the Sinus node (see Figs. 3.2 and 3.6). The

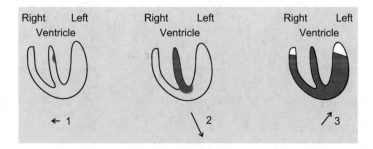

Fig. 3.6 Spread of excitation in the ventricles of the heart. Three different states of excitation are illustrated. The arrows represent the orientation and magnitude of the resulting dipole vector. Dark areas indicate excited sections

excitation front can be described by the sum of dipole vectors; the resulting vector is indicated in Fig. 3.6 by number 1. Later on, the excitation has spread to the tip of the heart, and the resulting vector is marked by 2. Because of the large mass of heart muscles the amplitude has increased. When nearly most of the heart is activated, the vector is again smaller (see number 3 in Fig. 3.6). The 3 vectors represent 3 different momentary conditions of the heart cycle starting from the resting ventricles, where the dipole momentum is zero. A vanishing dipole momentum we will also find in the completely activated heart.

The tip of the vector will cycle through a loop as indicated in Fig. 3.7. How will then, qualitatively, the potential difference look like that we can record between two points at the body surface?

A typical recording technique is to measure the potential at the left leg with respect to a so-called indifferent electrode composed of electrodes at the two arms. Figure 3.7a shows schematically the loop of the dipole in the body. The changes of potential difference that can then be detected between left leg and the indifferent electrode can be obtained by projection of the dipole vector on the direction of recording indicated by the dashed line. The resulting signal will consist of the three peaks labelled with QRS (Fig. 3.7b). In a normal ECG (see Fig. 3.1) these components, indeed, can be detected; in addition, waves labelled with P and T occur, which can be attributed to activation of the atria and repolarisation of the ventricles, respectively. One can easily imagine how deviations

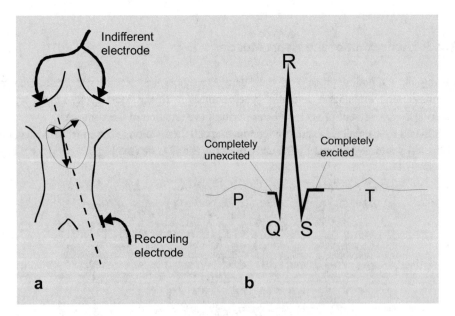

Fig. 3.7 Movement of the dipole vector in the body (**a**) and the resulting potential difference (**b**) detectable between left leg and the indifferent electrode (note that in reality the electrodes are connected at the wrists and the ankle, respectively)

from the normal ECG can be interpreted by irregularities in the spread of excitation. Figure 3.1 (lower trace) illustrates as an example schematically atrioventricular heart block showing the autonomic activity of the ventricular myocardium (governed by the AV node (see Fig. 3.2)) at slower rate than the occurrences of the P waves.

3.3   Recording Electrical Signals from Tissue

In this section we only briefly list three examples to illustrate in which sense recording from tissue can provide information for better understanding of electrophysiological processes.

3.3.1   Intracardiac Electrograms

More accurate information on the electrical characteristics of the beating heart, discussed in the previous section, can be obtained by electrodes that are brought into the heart via catheters. Such *intracardiac electrograms* reveal of course more details than the normal ECG (see schematic drawing in Fig. 3.8). We do not want to go into details here, but just take this as an illustration that direct contact between electrodes and the tissue gives additional information.

3.3.2   The Ussing Chamber

In context with recordings from tissue, we should at least briefly mention the classic method of the so-called *Ussing chamber* (Ussing & Zerahn, 1951), which allows to obtain

Fig. 3.8 Schematic drawing of a conventional ECG with P and QRS waves (**a**) compared to an intracardiac electrogram with further resolved P wave (**b**)

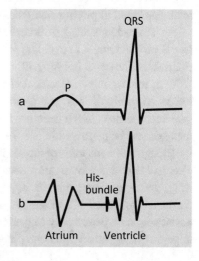

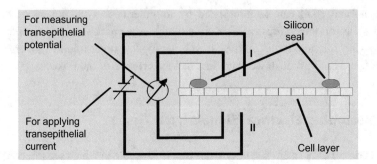

Fig. 3.9 The Ussing chamber with a piece of cell layer separating two compartments (I and II) by Silicon seals and electrodes for current stimulation and potential recording

basic information on the function of epithelial cell layers. The principle is shown in Fig. 3.9.

With this technique potential differences that exist across the epithelial layer can be measured. In addition, current pulses can be applied, and the resulting potential changes can be measured. An often-used procedure is to measure the current that is necessary to clamp the transepithelial potential to 0 mV. With identical solutions on both sides of the layer no holding current should be necessary. The persistence of a current (short circuit current) then reflects electrogenic, active transport.

3.3.3   Recording from the Brain

Like with intracardiac electrograms, more detailed information about the electrical activity of the brain can be obtained by implantation of electrodes into specific locations of the brain compared to placing electrodes on the skull.

For the understanding of brain function, the application of voltage-clamp techniques (see Sect. 3.4.5 and 4.1) to single cells in brain slices has become a powerful method. An example is illustrated in Fig. 3.10.

With this technique electrical signals from single cells can be recorded that are still imbedded within the neuronal network. With this procedure it became possible to investigate interactions between neurones in the network of even higher vertebrates. The relevant methods will be discussed in later sections (see e.g. Sect. 3.3.4).

Though it is not a topic of electrophysiology, we like to mention that the changes in the electrical neuronal activity are associated with the changes of magnetic fields in the range of fT, which are tiny compared, e.g. to the magnetic field in the μT range at the surface of the earth (see Sect. 9.1.1, *Magnetostatic fields*). Such changes in magnetic field can, nevertheless, be detected by Superconductance Quantum Interference Devices (SQUIDs) to perform magnetoencephalography (MEG).

Fig. 3.10 Neuron from periaqueductal grey area of rat brain attached to a patch pipette (from Shuanglai Ren and Wolfgang Schwarz, unpublished)

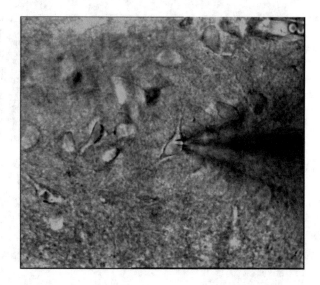

3.3.4   Recording Extracellular Field Potentials with Multielectrode Arrays

In addition to the various versions of voltage clamp (Sect. 4.1) there exists a method that allows recording of extracellular field potentials from electrically active cells or tissue. Extracellular field potentials can be detected whenever individual cells, cellular networks or cells in a tissue generate fast changes in their membrane potential. In most cases these changes are caused by action potentials. Extracellular field potentials around cells can be detected by capacitive coupling between cell membrane and an extracellular electrode close enough to the cell. Since electrical activity is a prerequisite for using this technique, brain or cardiac slices, neuronal or cardiac cultures, and retinal preparations are typical targets (Boven et al., 2006; Stett et al., 2003; Spira & Hai, 2013).

One of the major advantages of extracellular field potential recording is that this method enables non-invasive recording from cells or tissue, i.e. no electrodes penetrate the cell membrane. Therefore, culture dishes have been developed with embedded metal electrodes on their bottom. Tens or hundreds of electrodes are arranged in arrays with diameters of tens of microns at distances of tens to hundreds of microns (multielectrode array MEA). Together with highly developed amplifiers electrical signals can be recorded with high temporal and spatial resolution. Figure 3.11 shows a typical MEA used for recording from neuronal or cardiac cell cultures.

MEAs together with high-quality amplifiers and software are commercially available as ready to use systems and hundreds of publications have been published during the last years.

Recently, a microelectrode array system based on the CMOS technology has been developed with 4225 recording and 1024 stimulation electrodes (Stutzki et al., 2014). The principle of this system is based on the early work of the German scientist Peter

Fig. 3.11 MEA culture dish
with 64 recording electrodes and
one larger reference electrode.
The electrodes of this MEA have
a 10-micron diameter and are
arranged at a distance of
100 microns to each other
(provided by Multi Channel
Systems GmbH)

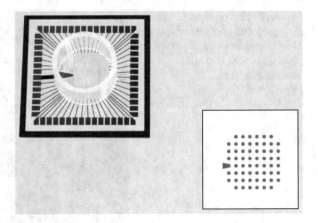

Fromherz et al. (1991) who started in the late 1980s trying to couple electrical nerve signals
to transistors.

3.4 Recording Electrical Signals from Single Cells

The most powerful electrophysiological method for basic research is the voltage-clamp
technique, which allows at a given membrane potential the measurement and analysis of
currents across the cell membrane that are mediated by specialised channels and carriers.
The analysis of voltage dependencies forms the basis for most electrophysiological
investigations. The voltage-clamp technique was prerequisite for the two milestones in
modern electrophysiology: the Hodgkin–Huxley description of excitability (see Sect. 6.
1.2) and the demonstration of single-channel events by Erwin Neher and Bert Sakmann
(see Sect. 4.4).

For detection of potential difference and current across a cell membrane microelectrodes
are impaled into the cell (Sect. 3.4.2). In electrophysiology ions are the current-generating
particles, in electronics, on the other hand, electrons produce the current. In electrophysiology, the coupling between the ionic and electronic world is usually established by using
Ag/AgCl electrodes (Sect. 3.4.2).

3.4.1 The Ag/AgCl Electrode

To measure potentials in solution, the Ag/AgCl electrode has become the standard system.
The principle of this electrode is illustrated in Fig. 3.12. Very often a silver wire electrolyt-
ically coated with silver chloride is used instead.

Fig. 3.12 Principle of Ag/AgCl electrode

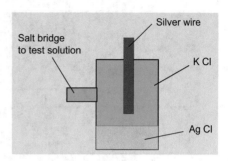

In the Ag/AgCl electrode system the charge carriers in the silver wire are the electrons e^-, in solution the chloride ion Cl^-. The electrode reaction will be governed by

$$AgCl + e^- \rightleftharpoons Ag^+ + Cl^- + e^- \rightleftharpoons Ag + Cl^-.$$

Due to the low solubility product K_L of AgCl ($a_{Ag^+} \cdot a_{Cl^-} = 1.7 \cdot 10^{-10}$), the KCl solution will be saturated, but will have a very low activity of Ag^+ (a_{Ag^+}). Against a standard hydrogen electrode, the Ag/AgCl electrode will show a potential difference of:

$$E_{el} = E_0 + \frac{RT}{F} \ln a_{Ag^+} = E_0 + \frac{RT}{F} (\ln K_L - \ln a_{Cl^-}) = E_0^* - \frac{RT}{F} \ln c_{Cl^-}$$

with $E_0^* \approx 0.222$ V under standard conditions (25 °C). The above equation shows that the Ag/AgCl electrode acts like a Cl^--selective electrode. This is an important characteristic with severe consequences if the surrounding Cl^- concentration is altered. Changes in Cl^- activity will result in the changes of the electrode potential, which will superimpose the potential difference measured between this and a second electrode.

3.4.2 The Microelectrode

To apply a voltage-clamp pulse to a cell, it is necessary to have intracellular electrodes. We will see later on that one possibility is to work with isolated cut pieces of cell fibres that allow access to the cytoplasm via the cut ends (Sects. 4.1.1 and 4.1.2). For working with intact cells, microelectrodes (first introduced by Ling & Gerard, 1949) must be used. They are pulled from glass capillaries ending up with tip diameters of less than 0.5 μm (see Fig. 3.13).

This allows penetration of the cell membrane without much harm to the cell. The electrical contact between cytoplasm and the electronics is achieved by filling the capillaries with an electrolyte solution and connecting via an AgCl-coated silver wire. As electrolyte solution a highly concentrated KCl solution (1–3 M) is often used. This brings the resistances of such pipettes down to the MΩ range and makes liquid-junction potentials

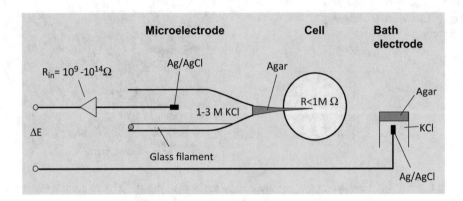

Fig. 3.13 Arrangement of a microelectrode (for details see text)

occurring at the tip of the electrode (due to the similar mobility of potassium and chloride ions) nearly independent of changes in the composition of the outer solution. As reference an extracellular bath electrode is used, also made of Ag/AgCl (see Fig. 3.13). A separation made by a so-called agar bridge (preventing solution flow but not charge flow) can be used to avoid direct contact between electrode solution and cytoplasm or bath, respectively.

The potential difference between the two electrodes is composed of several contributions:

$$\Delta E = \left(E_{\text{Mel}} + E_{\text{pip}} \right) + E_m - \left(E_{\text{bath}} + E_{\text{Bel}} \right) = \left(E_{\text{Mel}} - E_{\text{Bel}} \right) + \left(E_{\text{pip}} - E_{\text{bath}} \right) + E_m$$

with membrane potential E_m, potentials at the Ag/AgCl electrode of pipette and reference electrode E_{Mel} and E_{Bel}, respectively, and the potential between electrode and cytoplasm and bath, E_{pip} and E_{bath}, respectively. Since microelectrodes have resistances in the MΩ range, pre-amplifiers with input resistances of more than 10^9 Ω are used for recording the signal to avoid significant potential drops (see voltage follower, Sect. 3.4.5.3).

Before we will introduce the different variants of voltage clamp, we will give some background information about further applications of microelectrodes.

3.4.3   Ion-Selective Microelectrodes

3.4.3.1 Construction of Ion-Selective Microelectrodes

To determine the activities of ions in small volumes or even inside a cell, ion-selective microelectrodes have been designed (for details see e.g. Thomas, 1978). Because the ion-selective electrode is measuring the sum of the activity-dependent potential (see Sect. 3.4.3.2), which is generated at the tip of the electrode, and the resting potential of the cell, the resting potential has to be measured independently by a second reference electrode (Fig. 3.14). Then, the difference between the potentials of the two intracellular electrodes,

Fig. 3.14 Diagram for
measuring with ion-selective
microelectrodes. The membrane
potential measured by the
reference electrode is subtracted
from the signal of the
ion-selective electrode

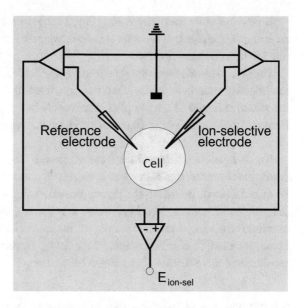

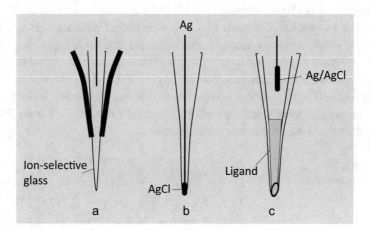

Fig. 3.15 Ion-selective microelectrodes. (**a**) Ion-selective glass, (**b**) solid-membrane, and (**c**) liquid-membrane electrodes

$E_{ion-sel}$, can be used to calculate the intracellular ion activity (see Fig. 3.14) based on calibration with suitable solutions.

Three variants of ion-selective microelectrodes are illustrated in Fig. 3.15.

Glass-microelectrodes (Fig. 3.15a) are made of ion-selective glasses. The basis of ion-selectivity will be discussed later (see Sect. 5.1.1).

Solid-membrane microelectrodes (Fig. 3.15b) have a hardly soluble salt at their tip. As an example, the Ag/AgCl version is shown, where the Ag/AgCl electrode is tightly sealed into the tip of a microelectrode.

The *liquid-membrane microelectrode* (Fig. 3.15c) is the most common version of ion-selective microelectrodes that can be used for intracellular measurements. An ion-selective ligand in a hydrophobic solvent is brought into the tip of a microelectrode. Because of the hydrophilic character of glass, the capillaries have to be silanised before filling.

Another problem arises since these electrodes have an extremely high electric resistance, and therefore, tips with wider openings than used for normal microelectrodes have to be manufactured. Bevelling the pipette tips and using double-barrelled electrodes reduce the problem of making two large holes into the cell membrane. In the double-barrelled electrodes the second channel serves for the intracellular electrode measuring the membrane potential of a cell (compare Fig. 3.14). Nevertheless, pre-amplifiers with input resistance of more than 10^{14} Ω have to be used.

3.4.3.2 Theory of Ion-Selective Microelectrodes

The principle of a liquid-membrane ion-selective microelectrode is that of an ion-exchanger membrane. The basis of the ion-exchanger membrane is a network of fixed charges within a solvent that contains moveable counter charges (see Fig. 3.16). We call the membrane cation exchanger if the fixed charge is negative (e.g. $-(SO_3^-)_n$), and anion exchanger if the fixed charge is positive (e.g. $-[N^+(CH_3)_3]_n$). As we have discussed previously for a cell membrane that is impermeable for one ion species (in the present case the fixed charges), a Donnan potential will be generated (see Sect. 2.3) at the interface of the bath solution and the unselective ion exchanger.

Fig. 3.16 Principle of ion-exchanger membrane

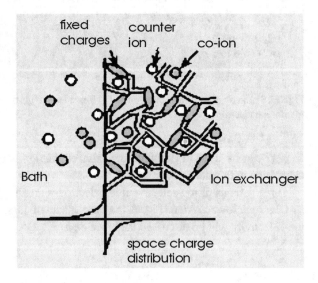

If the ion-exchanger membrane separates two electrolytes of different concentrations, an equilibrium will develop with vanishing electrochemical potential differences across the membrane:

$$\Delta G = \mu' - \mu'' = 0$$

with.

μ representing the electrochemical potential in compartment $'$ and $''$, and

$$\mu = \mu^0 + RT\ln a + zF\Phi$$

μ^0 chemical potential under standard conditions; a ion activity (with respect to 1 mol); z valency of the ion; Φ electrical potential.

For a single permeable ion species, the electrical potential difference across the ion-exchanger membrane, $\Delta\Phi = \Phi'' - \Phi'$, is

$$\Delta\Phi = \frac{RT}{zF} \ln \frac{a''}{a'} = E_{\text{Nernst}}.$$

For an ion-selective microelectrode the ion exchanger has to be composed of an ion-selective ligand. If one takes into account that the exchanger has a certain selectivity, the potential that is detected follows a semi-empirical equation developed by Nicolsky (see Thomas, 1978). For ions with the same valency this equation can be written in the form:

$$\Delta\Phi = \frac{RT}{zF} \ln \left(\frac{a''}{a'} + \sum_j k_j a_j \right)$$

where the subscript j refers to the other interfering ions at concentration (activity) a_j with relative distribution coefficients k_j.

3.4.4   The Carbon-Fibre Technique

Communication between cells is usually mediated via the release of transmitters or hormones by an effector cell; the molecules then reach their receptor cell via diffusion or transport in, e.g. the blood. To detect the amount of release or changes in the concentration at the surface of a single cell at sub-micromolar levels, carbon-fibre microelectrodes have been designed (Ponchon et al., 1979; Gilmartin & Hart, 1995) (see Fig. 3.17) for oxidisable or reducible molecules. At the tip of the microelectrode the molecules T can be oxidised or

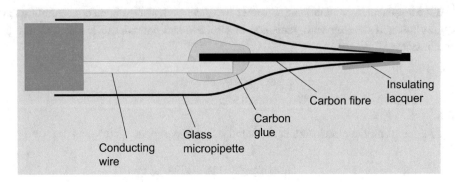

Fig. 3.17 Carbon-fibre microelectrode

reduced by application of a positive or negative voltage, respectively, according to the reaction:

$$T_{\mathrm{red}} \; T_{\mathrm{ox}} + n \cdot e^-.$$

The current that is associated with the oxidation or reduction process is a measure for the local concentration of the compound.

3.4.4.1 Construction of Carbon-Fibre Microelectrodes

For the construction of the microelectrodes, a single carbon fibre used as the conducting element is inserted into a glass capillary. One end of the carbon fibre is glued to a conducting wire leading to the preamplifier; the other end of the capillary is then pulled out to a fine microelectrode with a tip diameter of about 5 µm, the outer wall of which is covered by an insulating lacquer. For application of the technique an additional reference electrode is necessary, which can be a conventional microelectrode. The arrangement would be similar as for ion-selective microelectrode (Fig. 3.14).

3.4.4.2 Theory of Carbon-Fibre Microelectrodes

In a solution with physiological ionic strength, the potential at the tip of an electrode decays within a few nm. To detect changes in concentration of a compound, the tip of the electrode has to be brought into close proximity to the cell surface. The distribution between the oxidised and the reduced form of the compound T at the tip of the electrode at a potential E will be governed by Boltzmann distribution:

$$\frac{T_{\mathrm{ox}}}{T_{\mathrm{red}}} = e^{\frac{nF}{RT}(E-E_0)}$$

with E_0 the standard potential and n the number of transferred electrons. The current flow associated, e.g. with the transfer between the reduced and the oxidised form in response to a potential difference E–E_0 is given by the Butler–Volmer formula:

$$j = -k_e FT_{red} \text{ with } k_e = k_0 e^{(1-\alpha)nF}$$

where k_e is the electron transfer rate and α the transfer coefficient.

Taking into account that the concentration at the electrode surface differs from the bulk concentration (of course we should more correctly name it activity) and assuming a diffusion coefficient D_{red} and a diffusion layer thickness δ, the current can be described by

$$j = -\frac{k_e FT_{red}^{bulk}}{1 + \frac{k_e \delta}{D_{red}}}.$$

If the potential E is made large enough, the electron transfer rate k_e becomes large and hence:

$$j = -FT_{red}^{bulk} \frac{D_{red}}{\delta}.$$

3.4.4.3 Amperometric and Cyclic Voltammetric Measurements

To determine the concentration and identity of the molecule of interest, two different protocols can be applied: amperometry (see e.g. Gomez et al., 2002) and cyclic voltammetry (see e.g. Kawagoe et al., 1993).

In amperometry a constant potential is applied, large enough so that the above equation holds. Under these conditions the current is diffusion-limited and becomes voltage-independent. Changes in current with time are measured and reflect changes in the concentration. This method is often used to follow the release of transmitter substance from synaptic nerve endings (see e.g. Sect. 6.3.2).

In fast scan cyclic voltammetry, a cyclic potential ramp is applied to the measuring electrode (Fig. 3.18a) lasting a few ms. The current flowing across the electrode is measured before and after changes in the concentration of the substance to be detected (Fig. 3.18b), the small difference in current is typically in the nA range (Fig. 3.18c) and represents the oxidising and reducing current, respectively. Since the oxidising and reducing potentials are substance-specific, this method with varying potential can be applied for chemical identification.

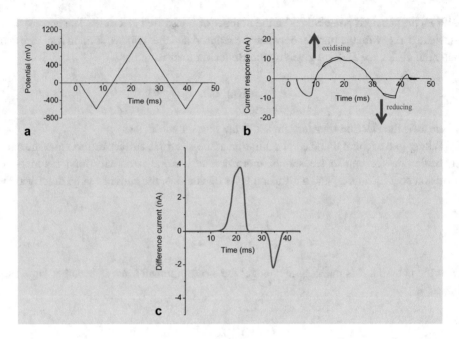

Fig. 3.18 Schematic potential protocol (**a**) and current response (**b**) in cyclic voltammetry. The tiny current differences (**c**) of a few nA represent oxidising and reducing current

3.4.5   Basics of Voltage Clamp

In the following we will start with the "Ideal Voltage Clamp" from which the "Real Voltage Clamp" will be developed step by step.

3.4.5.1 The Ideal Voltage Clamp

The ideal voltage clamp (Fig. 3.19) consists of a voltage source providing the clamp potential V_C, the model membrane (membrane resistance R_M and capacitance C_M in parallel), a switch, and an ampere meter for measuring membrane current I_M.

This circuit is "ideal" since wires, ampere meter and battery are assumed to have negligible internal resistance. Therefore, after the switch is closed, the model membrane reaches the potential of the battery as soon as the capacitance is charged ($V_M = V_C$).

3.4.5.2 The Real Voltage Clamp

The main difference between the ideal and the real voltage clamp is that the connection between the electronic circuit and the cell (in which currents are carried by ions) cannot be treated with negligible resistance. In many cases glass microelectrodes (Sect. 3.4.2) are used that have resistances R_E in the range of MΩ similar to the input resistance of large cells such as frog oocytes (see Sect. 7.1.2). Therefore, one has to add the electrode resistance to the "ideal" circuit shown in Fig. 3.19 ending up with Fig. 3.20. The problem that we are

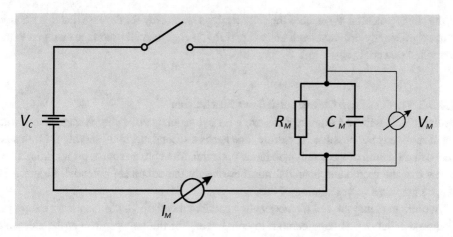

Fig. 3.19 The ideal voltage clamp

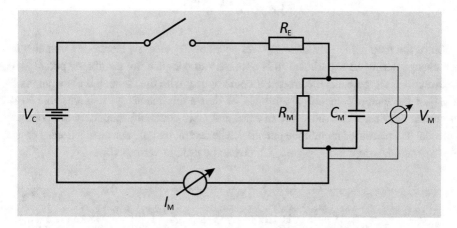

Fig. 3.20 The real voltage clamp

now confronted with is that two resistors in series act as a voltage divider with the characteristic:

$$V_M = V_C \frac{R_M}{R_M + R_E}.$$

This means that only a fraction

$$R_M/(R_M + R_E)$$

of the clamp potential V_C reaches the cell membrane, and only for the case that this fraction is not significantly different from unity (i.e. if $R_M \gg R_E$), the cell is voltage clamped to the command potential ($V_M \approx V_C$).

3.4.5.3 The Voltage Clamp with Two Electrodes

For large cells with low input resistances, it is obvious that performing voltage clamp with one electrode is not possible. Therefore, one needs a second electrode serving for independent determination of the actual membrane potential. The voltage source is then adjusted in a way that the membrane potential matches exactly the command potential. Figure 3.21 gives a graphical representation of this.

In order to clamp the cell membrane to a certain potential V_M, it is necessary to apply a clamp potential that is large enough to compensate for the voltage drop at the electrode resistance R_{CE}, which is quantitatively described by

$$V_C = V_M \frac{R_M + R_{CE}}{R_M}.$$

Since the membrane resistance R_M and occasionally also the electrode resistance R_{CE} can change during an experiment, it is necessary to compare the membrane potential V_M measured via the potential electrode PE continuously with the command potential, and to re-adjust the clamp potential V_C. Instead of doing this manually, it is possible to use electronic devices, which allow for an exact and rapid communication between command potential and measured membrane potential. The central part of such an electronic set-up is the "operational amplifier"(op-amp) which can be used in various ways.

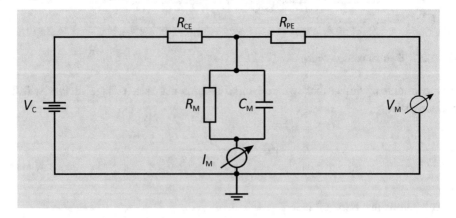

Fig. 3.21 Voltage-clamp circuit with current electrode (CE) and potential electrode (PE)

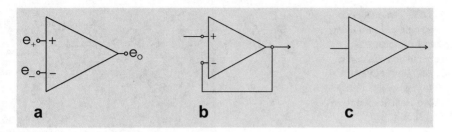

Fig. 3.22 Schematic drawing of an operational amplifier (**a**) and a voltage follower (op-amp with unity gain, **b**). (**c**) Simplified presentation of a voltage follower

The main characteristic of an op-amp (Fig. 3.22) is its ability to amplify the difference between its two inputs by a factor A (gain):

$$e_0 = A(e_+ - e_-).$$

When the negative input is connected to the output, the op-amp works as voltage follower (Fig. 3.22) meaning that the output signal equals the signal at the negative input:

$$e_0 = A(e_+ - e_-) = A(e_+ - e_0) \quad \Rightarrow \quad e_0 = \frac{A}{A+1}e_+ \approx e_+ \quad \text{mit } A = 10^4 - 10^6.$$

A simplified presentation of the voltage follower (Fig. 3.22) will occasionally be used in some drawings.

The two op-amp variants can be used to complete the two-electrode voltage-clamp circuit as shown in Fig. 3.23 , which forms the basis of most of the commercially available amplifiers.

The voltage follower (**VF**) is used to uncouple the sensitive signal of the voltage electrode from the following devices such as voltage-clamp amplifier, oscilloscope or pen recorder, and to serve as a high resistance input in order to minimise the current flow through the voltage electrode. The second op-amp is used as a negative feedback amplifier (**FBA**) with high gain for voltage clamping. The positive input is connected to the command potential, the negative input to the signal delivered by the voltage follower. These two input signals define the potential at the output, and hence allow the cell to be clamped fast and accurately to the command potential. The current flow from the feedback amplifier is identical to the membrane current and can be measured either at the output of the op-amp or at the grounded bath electrode. Very often two bath electrodes are used, one current-passing grounded electrode and one bath electrode serving as a reference electrode for the intracellular voltage electrode (virtual ground). The use of two bath electrodes has the advantage that only the grounded electrode passes large currents and the non-grounded, virtual-ground bath electrode cannot polarise due to current flow. Also, voltage errors

Fig. 3.23 Two-electrode
voltage-clamp circuit using
op-amps for measuring
membrane potential via a
voltage follower (VF) and
performing voltage clamp by a
negative feedback amplifier
(FBA) circuit

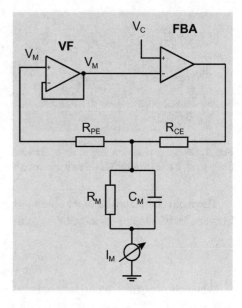

arising from series resistance of the bath medium, which is a problem if large currents are to
be measured, are minimised by the use of two bath electrodes.

3.4.5.4 One-Electrode Voltage Clamp Used for the Patch-Clamp Technique

In the previous paragraphs we learned about the impossibility to perform voltage clamp
with one electrode if the resistance of this electrode and the resistance of the cell membrane
under study are of similar magnitude. However, if the membrane has a significantly higher
resistance than the electrode, the difference between command and membrane potential
(i.e. the influence of the series resistance added by the electrode) becomes negligible. This
circumstance can be utilised for voltage clamping small cells or even small membrane
patches with only one microelectrode leading directly to the application of the patch-clamp
technique (see also Sect. 4.1.4). There are different variants of the patch-clamp technique
used that will be discussed in more detail in Sect. 4.4. However, all these variants have in
common that the voltage-clamped membrane, either from a whole cell or from a membrane
patch, has a resistance in the range of $G\Omega$.

Since typical patch microelectrodes have resistances between 0.5 and 50 $M\Omega$, voltage
control can be achieved with only one electrode that measures potential and passes current
simultaneously. Figure 3.24 gives a schematic representation of the minimum electronic
circuit.

One can derive the circuit for the patch-clamp method from that of the two-electrode
method simply by omitting the current electrode and connecting the output of the negative
feedback op-amp back to the "voltage electrode" via a resistor (feedback resistor). The

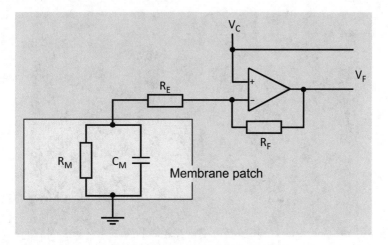

Fig. 3.24 Electronic circuit for the patch-clamp technique

membrane current I_M can now be calculated by measuring the voltage drop V_F across the feedback resistor R_F with

$$I_M = V_F/R_F.$$

Since the command potential equals the potential at the negative input (the measured membrane potential) of the op-amp during voltage clamp, the current measurement is usually performed by measuring the voltage between the positive input and the output of the op-amp (see Fig. 3.24).

3.4.5.5 Performing Voltage Clamp

The first voltage-clamp method was described by Cole in 1949. The basic principle is the feedback amplifier (see above and Fig. 3.25), which allows to apply voltage pulses to the membrane and measure the voltage- and time-dependent changes of current responses that flow across the membrane to the grounded bath, and hence represent the membrane current.

The most common way of doing voltage-clamp experiments is the application of rectangular voltage pulses. The advantage of rectangular voltage steps is that after brief transient capacitive current time-dependent changes in membrane conductances can be analysed. A typical schematic recording from an excitable membrane is illustrated in Fig. 3.26 , where the current is composed of ion-specific (I_{Na}, I_K), leak (I_l), and capacitive current (CdE/dt):

$$I = I_{Na} + I_K + I_l + C\frac{dE}{dt}.$$

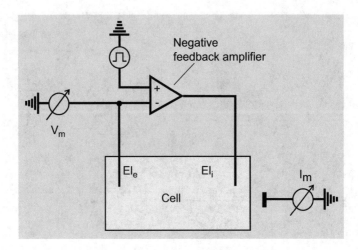

Fig. 3.25 Principle of voltage clamping a cell

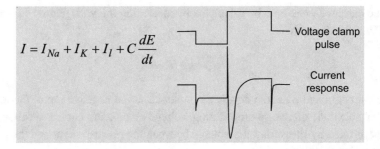

Fig. 3.26 Typical voltage-clamp recording (schematic)

As already mentioned in the introduction, an important step in Hodgkin's and Huxley's work was the separation of the ionic current components. Very useful in this sense are specific inhibitors. Figure 3.27 illustrates the action of tetrodotoxin, a very potent inhibitor of the Na^+ channels, which was extracted from the puffer fish.

3.4.6   Noise in Electrophysiological Measurements

Electrical measurements are often distorted by *noise*. In general, noise can be defined as any disturbance that adds to the measured signal of interest. In electrophysiological experiments such noise can arise from current fluctuations in the membrane, from electrodes, from the amplifier electronics or from external sources such as power lines, computers, monitors, and many other devices located in the periphery of the set-up (for

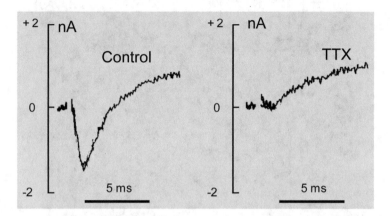

Fig. 3.27 The action of Tetrodotoxin (TTX) on membrane currents elicited by a rectangular voltage-clamp pulse (based on Hille (1970), Fig. 3, with kind permission from Elsevier AG 1970)

details see Axon Instruments, Sherman-Gold, 2008). Another source of noise can be the digitisation process (quantising noise, aliasing noise) if the filter settings are not optimal.

Most of the noise is of random nature, meaning that only the average value can be quantified, which is most commonly expressed as the root-mean-square (rms) value. If the noise has Gaussian distribution, the rms value means that there is a probability of 0.32 that the noise signal will exceed the rms value. Therefore, the peak-to-peak value of noise reaches approximately six times the rms value.

Total random noise E_T from several noise sources E_1, E_2, and E_3 adds on the basis of their rms values:

$$E_T = \sqrt{E_1^2 + E_2^2 + E_3^2}.$$

Depending on the source of noise frequency-dependent (often the so-called pink or flicker noise with $1/f$ dependency) and frequency-independent (white) noise can be detected (Fig. 3.28).

3.4.6.1 Thermal Noise

Thermal noise (also termed Johnson or Nyquist noise) is generated by the thermal motion of charged particles (electrons, ions) in a conductor. Thermal noise is equally distributed over all frequencies (white noise) and its spectral density S is given by

$$S = 4kTR_\Omega \quad \text{unit: V2/Hz}$$

where k is Boltzmann's constant, T is the temperature in K, and R_Ω is the resistance.

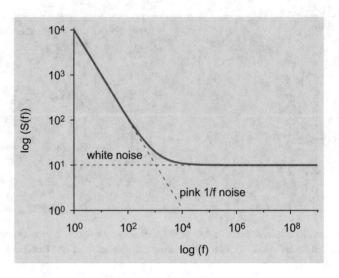

Fig. 3.28 Dependency of power spectral density S on frequency

The rms value E for a given bandwidth B is then:

$$E = \sqrt{4kTR_\Omega B} \quad \text{unit: Volt.}$$

3.4.6.2 Shot Noise
Shot noise is generated when the current flow occurs across a potential barrier as it is the case in transistors but not in simple resistors. The temperature and frequency-independent rms value of shot noise is given by

$$I_S = \sqrt{2qIB} \quad \text{unit: A}$$

where q is the charge of the elementary charge carrier, I the current flow through the noise source, and B the frequency bandwidth. Shot noise is usually negligible in electronic devices compared to the thermal Nyquist noise and frequency-dependent noise.

3.4.6.3 Dielectric Noise
Dielectric noise is thermal noise produced in capacitors and is dependent on the loss of the dielectric of the capacitor. The power spectral density of dielectric noise is

$$S_D = 4kTDC_D\omega \quad \text{unit: A2/Hz}$$

where ω is the frequency ($\omega = 2\pi f$) and D the dissipation factor of the capacitance C_D. The rms value for a given bandwidth B is

$$I_D = \sqrt{4kTDC_D\pi B^2} \quad \text{unit: A}$$

Dielectric noise in electrophysiological set-ups arises mainly from the electrode glass, but also high-loss dielectrics, which need not to be coupled directly to the electrode input (such as measuring chambers), can contribute significantly to this type of noise.

3.4.6.4 Digitisation Noise

Under optimum conditions digitisation noise is small compared to other noise sources and can, therefore, be neglected. Digitisation noise arises when the analogue current or voltage signal is transferred into a digital number and, therefore, is an integer multiple of the elementary quantity δ, the quantising step.

Very often 12-bit analogue-to-digital converters are used. When the full-range signal covers 10 V, the respective quantising step is

$$\delta = 10\text{V}/2^{12} = 2.44 \text{ mV}.$$

For a 16-bit converter one gets $\delta = 153$ μV, for 20-bit converter $\delta = 10$ μV. When δ is small compared to the full range, the rms value can be approximated by

$$E = \sqrt{\frac{\delta^2}{12}}.$$

During the digitisation the signal is not only quantised but also sampled at a certain frequency, the sampling frequency. If the signal is sampled every 10 μs, the sampling frequency is 100 kHz. According to sampling theorem, the quantising noise occurs in the frequency range between 0 Hz (DC-signal) and half the sampling frequency f. Within this frequency band the noise is equally distributed (white noise) with the spectral density:

$$S^2 = \frac{\delta^2}{6f}.$$

As mentioned earlier, the digitisation noise is not a severe problem when the parameters for digitisation are well adjusted. This means that δ should always be small compared to the measured signal. Under certain circumstances, for example, when the interesting signal is embedded in large background signal, digitisation noise becomes significant.

3.4.6.5 The Sampling Theorem and Aliasing Noise

According to the sampling theorem (see Lüke, 1999), a signal sampled with frequency f_s contains only frequency components smaller than $f_s/2$. This frequency $f_s/2$ is called Nyquist or folding frequency f_n. As a consequence, sampling of a signal, which contains frequencies higher than f_n, will lead to a loss of information that resides in the frequency domain above f_n, and more severe, will transfer noise from the frequency above f_n into the frequency band below f_n. The term f_n is called folding frequency, because the noise spectrum is folded around f_n. Quantitatively, this effect is described by

$$f_a = |\, f_i - af_s \,|\,,$$

where f_i is the frequency component above f_n, and a is a positive integer value which is chosen in a way that the alias frequency is folded in the frequency band below f_n. For example, if one samples a signal at 10 kHz, the folding frequency is 5 kHz (compare Fig. 3.29). If the analogue signal were not filtered, frequencies above f_n will fold into the sampled signal: At 2 kHz frequencies originally located at 8, 12, 18, 22, 28, etc. will appear. Since filters have no ideally sharp cut-off at their corner frequency f_c, the filter setting should be reduced below the value recommended by the sampling theorem to avoid aliasing noise. Under normal conditions it is useful to set the filter to $f_c = 0.4 - 0.5 f_n$.

3.4.6.6 Excess Noise

Excess noise is noise that originates from any noise source that cannot be classified into the noise classes already mentioned, and often displays $1/f$ dependency. Also signals arising from 50-Hz cycle power lines, mobile telephones, computer monitors, radio station, etc.

Fig. 3.29 Frequency artefact generated from a signal of frequency f_n if sampled at $0.5 f$

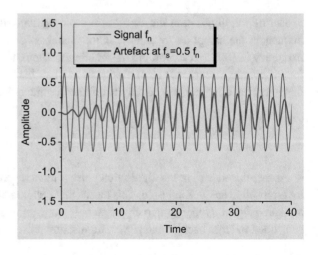

belong to excess noise. Particular attention has, therefore, to be paid to electrically shielding of the set-up.

Take-Home Messages
1. **Electromedical techniques** are predominantly applied for **diagnosis** (e.g. ECG), but also for **therapy**.
2. For high throughput screening **extracellular field potentials** can be measured with **Multielectrode Arrays**.
3. **The Ag/AgCl electrode** is a basic tool for recording electrophysiological signals. The electrode reaction can be described by

$$\text{AgCl} + e^- \rightleftharpoons \text{Ag} + \text{Cl}^- \qquad E_{el} = E_0^* - RT/F \ln c_{\text{Cl}^-}.$$

4. With **Ion-selective microelectrodes** respective ion activities can be detected even in small volumes (e.g. intracellularly).

   With **Carbon-fibre electrodes** oxidisable/reducible chemicals can be identified and quantified in small volumes (e.g. transmitter release).
5. **The Voltage-Clamp Technique** allows to clamp the membrane potential to defined values via **negative feedback amplifier (FBA)**.

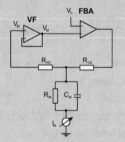

   A second electrode to measure the membrane potential via **voltage follower (VF)** is necessary if membrane resistance is not orders of magnitude larger than pipette resistance.
6. To avoid **Aliasing Noise** the cut-off frequency fc of the signal filter should be less than 0.5 of Nyquist frequency: **fc < 0.5 fn**.

Exercises

1. Describe the different phases of an electrocardiogram.
2. Name electromedical techniques that are still in use.
3. How would an ECG look like if two legs would serve the indifferent and the right arm as recording electrode?

4. Describe the principle of the Ussing chamber.
5. What is the meaning of short circuit current, and what can be concluded from this current?
6. How is a microelectrode array (MEA) composed?
7. What is the advantage of using MEAs compared to other electrophysiological methods on ells? What are disadvantages?
8. What kind of preparations can be used in MEA experiments?
9. What kind of electrical signals can be detected with MEAs?
10. What are the limitations of the MEA approach?
11. How does voltage-clamp work? What is the function of OP amplifier and voltage follower?
12. Under which conditions is it possible to perform voltage clamp with only one electrode?
13. What is the difference between voltage clamp and current clamp?
14. Describe a typical voltage-clamp protocol and argue the pulse protocol.
15. How is a microelectrode constructed? Which potentials have to be considered if one wants to determine a membrane potential?
16. What is the basis of an Ag/AgCl electrode?
17. What problems are associated with using an Ag/AgCl electrode as a bath electrode?
18. Why is 3 M KCl used in a microelectrode? Which problems may arise from using 3 M KCl?
19. Which ion composition should have the solution used in patch pipettes (Give an example)?
20. Formulate the sampling theorem and illustrate the consequences.
21. Which noise sources do you know?
22. How can noise be minimised?
23. How is an ion-selective microelectrode constructed?
24. What is the basis of an ion-selective electrode?
25. How is a carbon-fibre electrode constructed?
26. What can be measured with a carbon-fibre electrode?
27. What are the characteristics and advantages of a sniffer-patch electrode?

References

Boven, K.-H., Fejtl, M., Moller, A., Nisch, W., & Stett, A. (2006). On micro-electrode array revival. In M. Baudry & M. Taketani (Eds.), *Advances in network electrophysiology using multi-electrode arrays*. Springer Press.

Einthoven, W. (1925). The string galvanometer and the measurement of the action currents of the heart. *Nobel Lectures*, 1922–1941.

Fromherz, P., Offenhäusser, A., Vetter, T., & Weis, J. (1991). A neuron-silicon junction: A Retzius cell of the leech on an insulated-gate field-effect transistor. *Science, 252*, 1290–1293.

Gilmartin, M. A., & Hart, J. P. (1995). Sensing with chemically and biologically modified carbon electrodes. A review. *Analyst, 120*, 1029–1045.

Gomez, J. F., Brioso, M. A., Machado, J. D., Sanchez, J. L., & Borges, R. (2002). New approaches for analysis of amperometrical recordings. *Annals of the New York Academy of Sciences, 971*, 647–654.

Hille, B. (1970). Ionic channels in nerve membranes. *Progress in Biophysics and Molecular Biology, 21*, 1–32.

Katz, A. M. (1977). *Physiology of the heart*. Raven Press.

Kawagoe, K. T., Zimmerman, J. B., & Wightman, R. M. (1993). Principles of voltammetry and microelectrode surface states. *Journal of Neuroscience Methods, 48*, 225–240.

Ling, G., & Gerard, R. W. (1949). The normal membrane potential of frog sartorius fiber. *Journal of Cellular and Comparative Physiology, 34*, 383–396.

Lüke, H. D. (1999). The origins of the sampling theorem. *IEEE Communications Magazine, 37*(4), 106–108.

Ponchon, J., Cespuglio, L. R., Gonon, F., Jouvet, M., & Pujol, J. F. (1979). Normal pulse polarography with carbon fiber electrodes for in vitro and in vivo determination of catecholamines. *Analytical Chemistry, 51*, 1483–1486.

Sherman-Gold. (2008). *The axon guide for electrophysiology and biophysics laboratory techniques* (3rd ed.). Axon Instruments.

Spira, M. E., & Hai, A. (2013). Multi-electrode array technologies for neuroscience and cardiology. *Nature Nanotech, 8*, 83–94.

Stett, A., Egert, U., Guenther, E., Hofmann, F., Meyer, T., Nisch, W., & Haemmerle, H. (2003). Biological application of microelectrode arrays in drug discovery and basic research. *Analytical and Bioanalytical Chemistry, 377*, 486–495.

Stutzki, H., Leibig, C., Andreadaki, A., Fischer, D., & Zeck, G. (2014). Inflammatory stimulation preserves physiological properties of retinal ganglion cells after optic nerve injury. *Frontiers in Cellular Neuroscience, 8*, 38.

Thomas, R. C. (1978). *Ion-selective intracellular electrodes. How to make and use them*. Academic Press.

Ussing, H. H., & Zerahn, K. (1951). Active transport of sodium as the source of electric current in the short-circuited isolated frog skin. *Acta Physiologica Scandinavica, 23*, 110–127.

Application of the Voltage-Clamp Technique

4

Contents

Abstract

The basic method in electrophysiology is the voltage-clamp technique. Different ways of applying the voltage clamp shall be described in this chapter. Depending of the cell type, different versions of the conventional two-microelectrodes technique have been developed. In addition, the patch-clamp technique (a one-electrode voltage clamp) and examples for automated electrophysiology are explained.

Keywords

Two-electrode voltage clamp · Single-channel conductance · Patch clamp · Sniffer patch · Automated patch clamp

4.1   Different Versions of the Voltage-Clamp Technique

4.1.1   The Classic Squid Giant Axon

The first voltage-clamp setup was designed for application to the squid giant axon (Cole, 1949; Hodgkin et al., 1952) and is illustrated schematically in Fig. 4.1. The squid giant axon has a diameter of up to 1 mm and can be isolated over a length of several cm. A piece of nerve fibre is closed at one end and two wire electrodes are inserted from the other side for clamping the membrane potential. A section of this preparation is electrically isolated, and the current flowing across the membrane of this section to the grounded bath is measured.

An important feature of the giant axon is the possibility to change the composition of the internal medium. The cytoplasm can be squeezed out without harm to the membrane, and the axon can then be perfused with solution designed by the experimenter. The method, in general, is applicable to cell fibres of large diameter.

4.1.2   The Vaseline- or Sucrose-Gap Voltage Clamp

For thinner fibres insertion of electrodes becomes impossible. However, a piece of cell fibre with diameter down to 10 μm can be prepared, and a section of several tens of micrometre can be isolated electrically by Vaseline or sucrose gaps (Fig. 4.2). Electrical access of the two electrodes to the cytoplasm, necessary for voltage clamp, is achieved through the cut ends of the cell fibre (Stämpfli, 1954; Nonner, 1969).

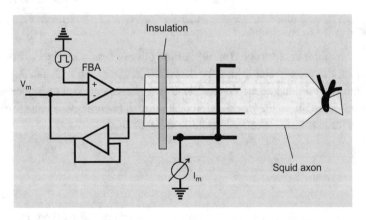

Fig. 4.1 Simple diagram of voltage clamping the squid giant axon

Fig. 4.2 Simple diagram of
voltage clamping thinner fibres
using Vaseline or sucrose-gap
insulation with electrical access
to the cytoplasm via the cut ends
of the fibre

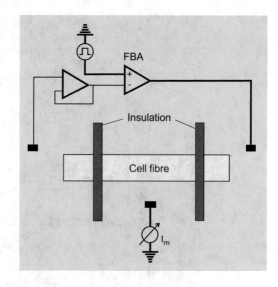

Fig. 4.3 Simple diagram of
voltage clamping ball-shaped
cells

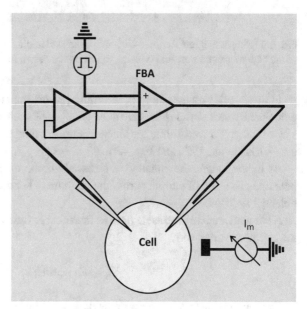

4.1.3   The Two-Microelectrode Voltage Clamp

If intact cells are to be used, the two-microelectrode technique can be applied (see Fig. 4.3).
This version of voltage clamp is predominately used for ball-shaped cells, and cells up to
more than 1 mm in diameter (like amphibian oocytes, see Sect. 7.1.2) can still be brought
under voltage clamp in the ms range if high-gain, high-voltage-clamp amplifiers are used.

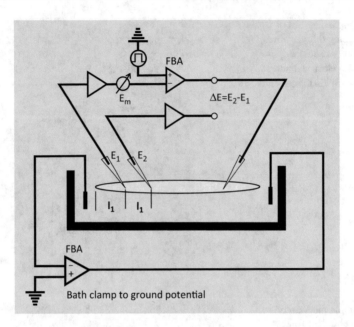

Fig. 4.4 Simple diagram for three-electrode voltage clamping a muscle fibre and with voltage clamp of the bath to ground potential (voltage followers are drawn in their simplified forms)

In these cells voltage clamp of the entire plasma membrane can be achieved. If, on the other hand, intact muscle fibres are to be voltage clamped, one runs into problems of proper space clamp. To avoid this, the three-electrode voltage clamp had been developed (see e.g. Adrian et al., 1970 and Fig. 4.4).

At E_1 the membrane potential is clamped via the feedback amplifier to the command potential. The loss of current across the membrane between E_2 and E_1 is measured as the potential drop ΔE.

If we treat a nerve or muscle fibre as linear cable (see Taylor, 1963), the potential along the cell fibre varies with

$$E_1 = E_0 \, cosh(l/\lambda).$$

The length constant λ is for a skeletal muscle fibre in the range of 0.1–1 mm.

When the microelectrodes are placed as indicated in Fig. 4.4, the current per length unit is described by

$$i_m(l_1) = \frac{E_2 - E_1}{r_i l^2} \, \frac{l_1^2 \cosh\left(l_1/\lambda\right)}{\lambda^2 \left[\cosh\left(2l_1/\lambda\right) - \cosh\left(l_1/\lambda\right)\right]}$$

with cytoplasmic resistance r_i.

Fig. 4.5 Simple diagram of voltage clamping ball-shaped cells with a single electrode

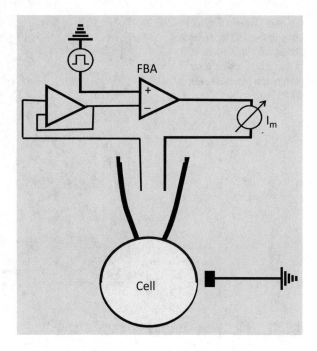

For $l_1/\lambda < 2$ we obtain

$$i_m = \frac{2}{3}\frac{E_2 - E_1}{r_i l_1^2}$$

with an error of less than 5% for l_1 of about 100 μm. The bath clamp is used to ensure that the entire bath is clamped at ground potential (see Fig. 4.4).

4.1.4   The One-Electrode Voltage Clamp

For small ball-shaped cells a simplified version of the two-electrode voltage clamp can be applied (see Fig. 4.5), where a single microelectrode with a wide opening, up to 50 μm in diameter, is placed on the surface of the cell (Kostuk & Krishtal, 1984); seal resistances of more than 10 MΩ can be obtained. After perforation of the membrane within the pipette opening, one obtains low resistance access to the cell interior, and voltage clamp can be achieved via two electrodes inside the same pipette.

Fig. 4.6 Simple diagram of the open-oocyte voltage clamp (based on Taglialatela et al. (1992, Fig. 1), with kind permission from Elsevier AG, 1992)

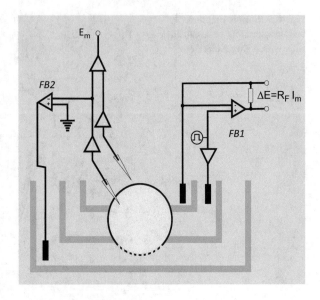

4.1.5   The Open-Oocyte Voltage Clamp

Since amphibian oocytes became very popular for electrophysiological investigations (see Sect. 7.1.2), a special version of the voltage clamp, the open-oocyte voltage clamp, has been designed as illustrated in Fig. 4.6 (Taglialatela et al., 1992). Main advantages of this voltage-clamp version are (1) reduced current noise, (2) control of the ionic composition of both the internal and external media, and (3) better time resolution. The latter has been used successfully to resolve and analyse ion channel gating currents (see also Sect. 4.3). The oocyte is placed into a chamber with three compartments that separate three sections of the oocyte by Vaseline seals. The feedback amplifier *FB1* clamps the outside of the upper membrane section to the command potential which is also applied to middle section. This minimises current flow via the seal between the two compartments. The lower section is perforated, and the feedback amplifier *FB2* clamps the inside to ground. The membrane current at clamp potential can be measured as voltage drop ΔE across R_F (Fig. 4.6).

4.2   Analysing Current Fluctuations

Before the introduction of patch-clamp technique into electrophysiology, information about single-channel characteristics could be extracted from current fluctuations originating from the superimposition of many single-channel events assuming that a channel can exist only in either an open, current-conducting or a closed state (see Fig. 4.7) .

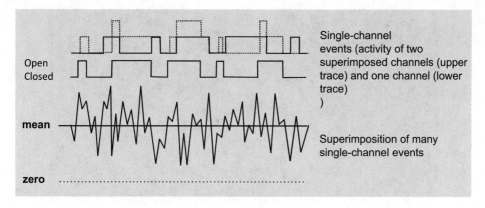

Fig. 4.7 Activity of single-channel events (upper two traces) leading to current fluctuations (lower trace). Upward deflections in the single-channel traces represent channel openings

The average current I of N channels with single-channel current i and open probability p_o is given by

$$I = N\, i\, p_o,$$

and the variance by

$$\text{var} = I \cdot i\,(1 - p_o) = N\, i^2\, p_o(1 - p_o) \text{ or } i = \frac{\text{var}}{I(1 - p_0)}.$$

By experimentally determining I, var and p_o one can calculate the number of involved channels and the single-channel current.

Another strategy is to perform a Fourier transform of the current fluctuations. With the assumption of only two states, the power spectral density is described by Lorentzian spectrum:

$$S_L = \frac{S_o}{1 + (f/f_c)^2}, \text{with } f_c = \frac{k_1 + k_{-1}}{2\pi}$$

where k_1 and k_{-1} are the forward and backward transition rates.

If the gating is governed by several mechanisms in series, the power spectral density can be described by the sum of Lorentzian components. Figure 4.8a shows as an example the superimposition of two Lorentzian components representing the fast activation and slower inactivation of the Na^+ channels of excitable membranes (see Sect. 6.1.2, The Na^+ Channel).

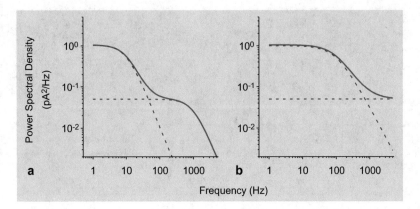

Fig. 4.8 Spectral densities of Lorentz type (**a**) with superimposition to two components with corner frequencies of 10 and 1000 Hz, and of diffusion type (**b**) with superposition of diffusion spectrum with corner frequency of 100 Hz and white noise (compared also Schwarz, 1983)

In case of an infinite number of open and closed states, a diffusion spectrum is obtained:

$$S_D = \frac{S_0}{1 + (f/f_c)^{\frac{3}{2}}}.$$

The fluctuations originating from the gating of K^+ channels can be described by a diffusion spectrum superimposed on a frequency-independent background (Fig. 4.8b).

4.3   Analysing Transient Charge Movements (Gating Currents)

Transport across the membrane by a carrier or opening and closing of a channel is always accompanied by conformational changes of the transport protein. If these changes are coupled to charge movements within the protein, the transition rates should depend on membrane potential. In the case of a channel with one open and one closed state, the distribution between the two states can be described by the Fermi equation

$$q = \frac{1}{1 + e^{-zEF/(RT)}},$$

which reflects the distribution of the movable charges between two states (see Fig. 4.9).

The charge movements can be detected as transient displacement currents, and the amount of moved charges in response to a voltage pulse is obtained by integration of the transient signal. From the steepness of the charge distribution versus membrane potential the effective valency z of charge moved per transport protein can be calculated. The z value reflects the total amount of charge moved in the electrical field times the fraction of the field

Fig. 4.9 Voltage dependence of charge distribution. The voltage dependency can be described by *a* Fermi distribution between two states. $z \cdot e^+$ represents the effective charge moved during the transition between the two states

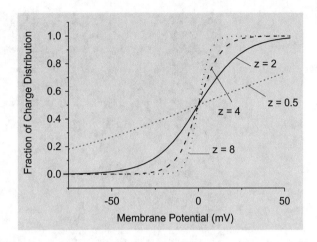

sensed by the moved charge and determines the steepness of the curve (see Fig. 4.9). From the total charge Q that can be moved the number N of involved channels or transporters can be calculated according to

$$Q = N z e^-.$$

The transient signal not only contains information about the amount and valency of the charges. In addition, the time course of the transient current can be described by exponential functions, the rate of which reflect transition rates between different conformations (see e.g. Sect. 7.1.4).

4.4   The Patch-Clamp Technique

The patch-clamp technique (Nobel Prizes; Neher, 1991; Sakmann, 1991) is a further development of the one-electrode voltage clamp (for details see Sakmann & Neher, 1995). The essential characteristic is an extremely high resistance formed by close contact between the pipette rim and cell surface reaching several GΩ (comp. Figure 4.10). The high seal resistance allows to resolve membrane currents in the pA range. This means that it is possible to record small currents across the membrane of small cells and even the current through a single open channel protein. To achieve the high seal resistance, several precautions related to pipette size and cleanness of surfaces and solutions have to be considered. The pipette should have an opening with smooth clean rim (see Fig. 4.11a). If such a pipette is placed onto the surface of a clean cell (cultured cells, enzymatically cleaned cells) in filtered solutions, and if slight negative pressure is applied to the pipette interior, seal resistances between 1 and 100 GΩ can spontaneously form. This high seal resistance is achieved by direct interactions between the surface of the glass pipette and the cell membrane on atomic dimensions. Most likely interactions are salt bridges between

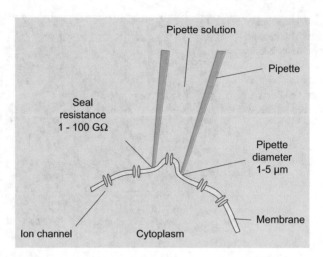

Fig. 4.10 Symbolised pipette attached to a cell membrane with a single channel in the patched membrane

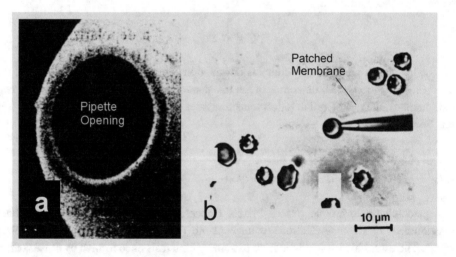

Fig. 4.11 Pipette tip of a patch electrode (**a**) and pipette touching a human erythrocyte with a piece of cell membrane sucked into the pipette (**b**) (based on Schwarz et al., 1989, Fig. 1, with kind permission from Elsevier AG, 1989)

negative charges on the glass and the membrane surface mediated by divalent cations and hydrogen bonds between O-groups on the glass surface and of the phospholipids forming the membrane as well as Van-der-Waals interactions.

Depending on pipette size, cell type, and channel density, the electrically isolated membrane patch can contain one or several channel proteins. Opening and closing of the channels results in sudden current changes that can be recorded under voltage clamp

Fig. 4.12 Simple diagram for patch clamp with feedback amplifier (FBA)

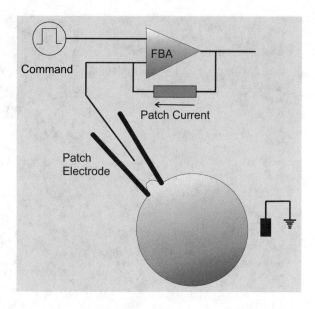

(Figs. 4.14 and 4.15). The principle of patch clamp (Fig. 4.12) is similar to the one-electrode voltage clamp (compare Fig. 3.24, Fig. 4.5).

4.4.1 Different Versions of Patch Clamp (Patch Conformations)

Due to the high seal resistance, the membrane-glass interaction exhibits also high mechanical stability. This allows to work with completely isolated membrane patches. The way of forming the different conformations, after a giga-ohm seal is achieved (cell-attached conformation), is illustrated in Fig. 4.13.

Inside-out membrane patches can be formed in different ways: After formation of the $G\Omega$ seal, the pipette can be withdrawn from the cell, and in the absence of Ca^{2+} (absence of divalent cations destabilises the integrity of the membrane) in the bath medium an isolated membrane patch can be obtained (Fig. 4.13a). In the presence of Ca^{2+} a vesicle is formed (Fig. 4.13c), but the surface oriented to the bath can be disrupted mechanically or chemically. This is also possible with the entire cell (Fig. 4.13b). For formation of an outside-out membrane patch, the patched membrane is first disrupted by slight negative pressure in the pipette or by application of a strong voltage-clamp pulse or the membrane may also be perforated chemically (Fig. 4.13d). This leads to the whole-cell configuration. After withdrawal of the pipette in Ca^{2+}-containing medium an outside-out patch is formed (Fig. 4.13e). A prerequisite for these pipette-withdrawal manipulations is that the cells are sticking to the bottom of the chamber.

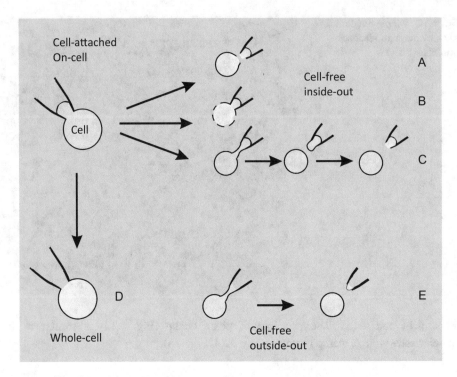

Fig. 4.13 Establishment of different conformations of the membrane patch

4.4.2   Advantages of the Different Patch Conformations

The characteristics of the various patch conformation are summarised in Table 4.1.

1. Cell-attached: Investigation of single-channel characteristics under normal physiological conditions. The pipette solution mimics external ion composition (solution exchange is in principle possible via special pipette holders). A disadvantage is that the exact potential across the cell-attached membrane is not known due to superimposition of the resting potential of the still intact cell on the clamped potential (compare Fig. 4.12).

2. Inside-out: Access to internal membrane surface via bath solution.
   Example: Ca^{2+}-activated K^+ channel in erythrocytes (Fig. 4.14).
   A problem may arise from an unknown, but essential internal substance that may be missing in the bath solution.

3. Outside-out: Access to external membrane surface via bath solution.
   Example: Inhibition of Cl^-- selective channels in K562 cells by H_2DIDS (Fig. 4.15).
   A problem may again arise with this configuration that an essential and unknown intracellular substance is missing in the pipette solution.

Table 4.1 Characteristics of different patch conformations

Conformation	Pipette solution	Bath solution	Characteristics	Multi-channel	Single channel	Carriers
On-cell	External		Physiological conditions Problem: E_m superimposed	(Yes)	Yes	No
Inside-out (e.g.Ca-activated channels)	External	Internal	Internal solution change (Ca, second messengers)	(Yes)	Yes	No
Outside-out (e.g. external inhibition)	Internal	External	External solution change (transmitter, drugs, toxins)	(Yes)	Yes	No
Whole-cell (e.g. carriers or low-cond. Channels)	Internal	External	External solution change Nystatin perforation	Yes	(Yes)	Yes
Giant-patch	Depending on mode		All variants possible Except for large whole-cell	Yes	(Yes)	Yes

4. Whole-cell: Behaviour of the complete cell with all its cytoplasmic contents can be studied if the membrane in the patch pipette is made permeable only for small ions either by, e.g., nystatin (perforated patch, which avoids loss of essential cytoplasmic components) or complete disruption. Partial exchange of cytoplasmic solution is possible via pipette solution. Only small cells can be used; but this technique also allows to measure currents generated by carriers or macroscopic ion channel currents (superimposition of gating of a large number of channels (Fig. 4.16).

5. Giant-patch: Conventional patch pipettes have tip diameter of 1–2 µm. To analyse multi-channel phenomena (like gating currents, see Sect. 4.3), the pipette diameter can be increased to 5 µm (macro patch). With special treatment of the pipette tip (see e.g. Rettinger et al., 1994) giga-ohm seals can be achieved with pipettes of diameters up to even 50 µm (giant-patch). This technique allows to detect currents generated by ion pumps and other carriers in excised membrane patches. Cell-attached, inside-out as well as outside-out configuration can be established with giant-patch pipettes.

In the following we briefly like to discuss, which information we can extract from conventional patch-clamp recordings.

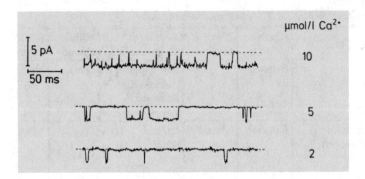

Fig. 4.14 Single-channel events in human erythrocytes (inside-out patch). Downward deflections represent channel openings (based on Schwarz et al., 1989, Fig. 5, with kind permission from Elsevier AG, 1989)

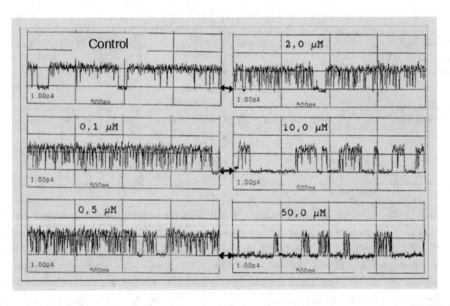

Fig. 4.15 Inhibition of Cl⁻ -selective channels in K562 cells by different concentrations of H_2DIDS in the bath medium (outside-out patch). Upward deflections represent channel openings

4.4.3   The Single-Channel Current and Conductance

Let us make a simple consideration to estimate the order of magnitude of single-channel currents and conductances using simple macroscopic rules for describing diffusion through a pore (see also Hille, 1992). The result, however, should be taken only as a rough estimate since it is of course questionable to apply macroscopic formalisms to microscopic (molecular) phenomena.

Fig. 4.16 Whole-cell recording of rat P2X$_1$ receptor that transiently opens as a cation-selective channel on extracellular application of ATP (Jürgen Rettinger, unpublished)

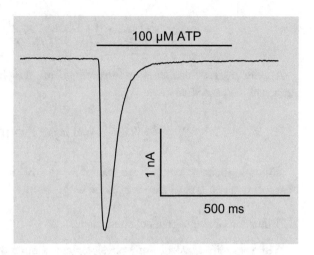

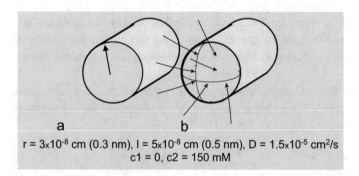

r = 3x10^{-8} cm (0.3 nm), l = 5x10^{-8} cm (0.5 nm), D = 1.5x10^{-5} cm^2/s
c1 = 0, c2 = 150 mM

Fig. 4.17 Geometry of a hypothetical pore without (**a**) and with (**b**) access resistance. For calculation of single-channel conductance the listed values for the radius r, length l, diffusion coefficient D and ion activities c1 and c2 were used (see also Hille, 1992, 2001)

1. Estimation of single-channel current:

Let us assume a cylindrical short pore with the parameters as illustrated in Fig. 4.17a. The current through such a pore would be

$$J = -zFAD\frac{dc}{dx} = -zF\pi r^2 D\frac{c}{l}$$

if there were no additional electrical potential gradient ($E_m = 0$). With the parameter given in Fig. 4.17 we obtain.

$J = 1.3 \cdot 10^{-16}$ mol/s or $7.7 \cdot 10^7$ ions/s corresponding to 12 pA.

Taking into account that access to the pore entrance may be restricted due to diffusion processes, an apparent increase in length of the pore may be considered. An access resistance can be added (Fig. 4.17b), which leads to:

$$R = \rho \left(\frac{l}{\pi r^2} + 2 \int_a^{\infty} \frac{\mathrm{d}r}{2\pi r^2} \right) = \frac{\rho}{\pi r^2} (l + r).$$

A more precise calculation by integrating from the planar rather than spherical entry surface of the pore yields:

$$R = \frac{\rho}{\pi r^2} \left(l + \frac{\pi r}{2} \right) \text{ and hence } l^* = (l + \pi r/2).$$

With this apparent length l^* we obtain: $J = 4 \cdot 10^7$ ions/s or 6 pA (remember that the transport rate of a typical carrier is in the range of only 1–100 translocations per second).

2. Estimation of single-channel conductance:

With a specific resistance of a physiological solution of $\rho = 100$ Ωcm (Table 2.2), we obtain a single-channel resistance of $R = \rho \, l/\pi \, r^2 = 1.8$ GΩ. Taking again into account the access resistance, we obtain $R = 3.5$ GΩ or $\gamma = 300$ pS. This would be the conductance of a wide, short pore that allows free diffusion of the ions across the membrane, and hence should represent a maximum single-channel conductance. Measured values are given in Fig. 4.18.

Recording single-channel currents provides information about the single-channel conductance and about the gating mechanism, which describes the transitions between open

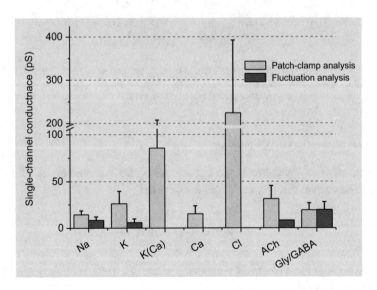

Fig. 4.18 Measured single-channel conductances of Na$^+$, K$^+$, Ca^{2+}, and Cl$^-$ channels and transmitter-activated channels data represent means + SD from different tissue and are based on data presented by Hille (1992)). Grey bars are from analysis by patch clamp, blue bars from fluctuation analysis

and closed channel conformations. Information we can obtain from these parameters are as follows:

1. The single-channel conductance:
(a) Classification of different channel types (compare Fig. 4.18) .
(b) Changes in the conductance can provide information about the mechanism of drug action.
(c) Changes in the conductance that result from mutations can provide information about structure–function relationships and physiological dysfunction.
2. The gating mechanism:
     For analysing gating properties, histograms can be calculated for the dwell times of a channel staying in an open or closed state. Because of underlying Markovian statistics, the histograms can be described by exponentials:

$$N = \sum A_i e^{-t/\tau_i}.$$

Each time constant τ_i can be interpreted as a mean dwell time of a channel in a particular state. A decrease in the mean dwell time for the open (τ_o) or closed (τ_c) state would mean an increased rate of leaving the corresponding channel state.

An example is shown in Fig. 4.19. Open-time and closed-time distributions in the absence of vanadate are described by a single exponential $(\tau_o \sim \tau_c \sim 5\ \mathrm{ms})$, in the presence of vanadate τ_o is not changed, closed-time distributions have to be described by two time constants $(\tau_{c1} \sim 1\ \mathrm{ms}, \tau_{c2} \sim 10\ \mathrm{ms})$. The number of time constants tells us about the number of open or closed states, respectively. The erythrocyte channel obviously has one open and one closed channel state, but in the presence of vanadate two closed states.

The histograms can give us information about:

1. Reaction diagrams with possible transitions between different channel states.
2. Mechanisms of action of second messengers, drugs, and toxins.
3. Structure–function relationships if mutants or chemically modified channels are compared with wild-type channels.

The distribution of the open and closed times can also be represented by a histogram with logarithmic bins for the probability-density (pd) function (Sigworth & Sine, 1987, and see Sakmann & Neher, 1995). This presentation forms a convenient way to determine the mean dwell times τ_i which appear in the pd function as maxima at $t = \tau_i$ (see Fig. 4.20). In this example the closed-time distribution could be described by the sum of four components, the open-time distribution by a single component.

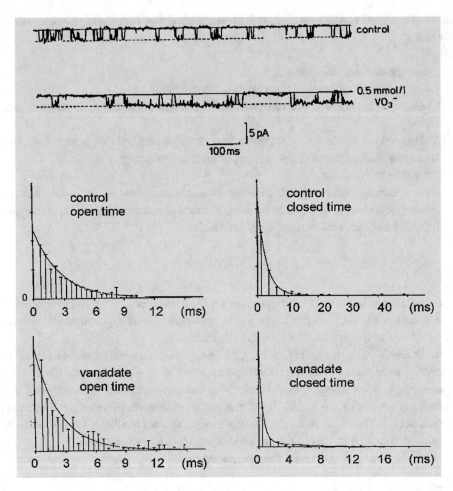

Fig. 4.19 Single-channel recording and histograms of open and closed times for Ca-activated K channel in human erythrocytes, and their modifications by application of vanadate (adapted from after Fuhrmann et al. (1985)). Note the different time scales for closed times. Downward deflections represent channel openings (based on Fuhrmann et al. (1985, Fig. 11), with kind permission from Elsevier AG, 1985)

4.4.4   The Sniffer-Patch Method

In Sect. 3.4.4 we have introduced the use of carbon fibre to detect electrochemically tiny amounts of a drug via redox reaction. The amperometric and cyclic voltammetric methods were described (Sect. 3.4.4.3), which allow fast and highly sensitive detection of molecules released as transmitters. Due to the presence of high levels of antioxidants under physiological conditions, usually positive potentials at the tip of the carbon fibre will oxidise the respective mediator. Premise for application of this method is that the molecule can easily be transferred between oxidised and reduced form at the tip of the carbon fibre.

The high sensitivity of the patch-clamp technique, on the other hand, allows using membrane patches of cells with over-expressed receptors that are highly specific to detect

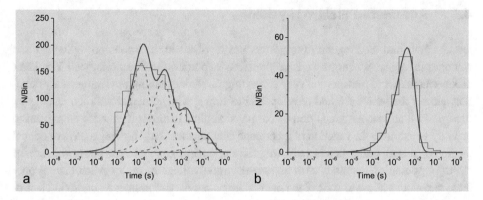

Fig. 4.20 Probability-density function of closed (**a**) and open (**b**) times of an anion-selective channel of K562 cells, an erythroid cell line (compare also Rettinger & Schwarz, 1994)

Fig. 4.21 Whole-cell current of a human mast cell (HMC-1) at +100 mV during superfusion of the cell with microdialysate from a rat that received during the experiment for 5 min acupuncture indicated by the grey bar (based on data provided by Meng Huang and Yong Wu with kind permission) (see also Wu et al., 2019)

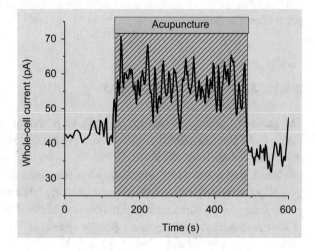

the interaction of the receptor protein with respective hormones, transmitters or other extracellular mediators released from another cell. With the so-called sniffer-patch method (see e.g. Allen, 1997; Muller-Chretien, 2014) the detection of a few molecules is possible using the response of ligand-gated ion channels in membrane patches. Calibrated current signals originating from activation of the "detector ion channels" yield highly sensitive and rapid responses. For signal detection the outside-out membrane patch or the whole-cell configuration can be used. The electrical response may be detected next to the releasing cell or at another location. Though this method has high spatial and temporal resolution, the range of concentration is limited to concentrations around the apparent K_m value of the receptor channel.

Release of mediators from cultured cells or from cells in tissue slices or light-induced release of caged molecules can be detected, and even changes of mediators in microdialysates may be monitored. Figure 4.21 illustrates as an example the detection of release of mediator by analysing microdialysate of rat under acupuncture treatment.

4.5   Automated Electrophysiology

Ion channels and electrogenic transporters are involved in numerous physiological and pathophysiological processes and are therefore important therapeutic targets. The latter makes channels and transporters very interesting for understanding physiological dysfunction and for the search for and development of drugs, which interact with ion channel or transporter function. Although non-electrophysiological methods have been developed to quantify or examine the function of membrane conductances (e.g. binding assays or optical methods), a real understanding of drug action on ion channel still needs the use of electrophysiological methods such as two-electrode voltage clamp or patch clamp; only these techniques allow to control the membrane potential. Unfortunately, classical electrophysiological methods are technically challenging, need well trained personnel, and generate results at very low throughput. This is in fact a serious problem in face of the large compound libraries that have to be scanned for lead compounds at the beginning of drug development. Therefore, robotic machines have been developed to automate the patch-clamp technique.

4.5.1   Automated Patch Clamp

The patch-clamp method has been developed in the late 1970 by the German scientists Neher and Sakmann (1976), who received the Nobel Prize in medicine in 1991 for their invention (Neher, 1991; Sakmann, 1991). Soon after the first papers using this method were published, first patch-clamp amplifiers were commercially available. The method became very popular, and almost every electrophysiology laboratory was equipped with a patch-clamp setup. Along with improvements in molecular and cell biology and the establishment of mammalian expression systems electrophysiology became more and more the bottleneck for screening of mutant ion channels in basic and pathophysiological research or for drug-screening in pharmaceutical industry.

Finally, the problem of the patch-clamp technique of ultra-low throughput, high cost per data point and need for skilled personnel was solved by the development of robotic systems that enabled automation of most if not all of the time consuming and technically difficult steps of manual patch clamping (see also Sect. 3.6).

When doing manual patching the experiment starts with preparing the patch pipette (pulling, heat-polishing the tip, filling the patch pipette, and inserting it into the pipette holder), finding the "right" cell under the microscope, moving the pipette tip (with positive pressure inside) close to the cell, touching the cell membrane, and applying negative pressure until the giga-ohm-seal forms. Then, the right configuration has to be established either whole-cell configuration, inside-out or outside-out. The experiment itself will be then a sequence of solution exchanges and/or voltage protocols in order to characterise the ion channel or transporter under the influence of drugs or modulators.

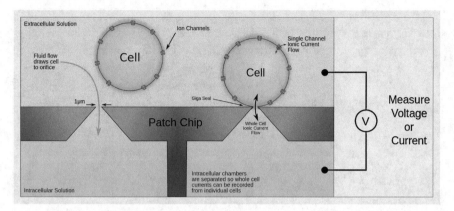

Fig. 4.22 Schematic drawing of a planar patch chip with a cell before approaching the chip opening and after establishment of whole-cell configuration (from https://en.wikipedia.org/wiki/Automated_ patch_clamp#/media/File:Patch_Clamp_Chip.svg)

Most of the robotic systems on the market automate the above listed manual steps by using planar chip substrates that contain one or more 1–2 μm openings (Fig. 4.22) mimicking the micrometre openings of classical patch pipettes (Fertig et al., 2002). Cells are transferred automatically into the wells either by microfluidics or by pipette robots and pressure differences to attract the cells, form the seal and establish whole-cell recording are accomplished by feedback-controlled pressure systems. Solution exchange is realised either by pipetting the solutions sequentially into the wells or by using microfluidic channels integrated into the multi-well chip plate.

This planar approach has two major advantages over manual patch clamp: automation, parallelization, and ease of use. With the newest systems on the market it is now possible to record from up to 768 cells in parallel fulfilling the demand of pharmaceutical industry to use the "gold-standard" patch-clamp technique already in earlier stages of drug development.

Take-Home Messages
1. Depending on the cell type **different versions of voltage-clamp** methods are in use:
    (a) Different versions for cell fibres (nerve and muscle fibres).
    (b) Different versions for more ball-shaped cells
        (cell diameter: 100 μm–1 mm).
2. **Current fluctuations can be analysed** to extract electrophysiological single-protein characteristics including:
    (a) Single-channel conductance.
    (b) Transport rates.
    (c) Number of proteins contributing to the signal.
3. **Transient charge movements can be analysed** to gain information about channel gating and partial reactions of carrier systems:
    (a) Effective valency of moved charges.
    (b) Number of proteins contributing to the signal.
4. Prerequisite for applying the **patch-clamp method** is:

$$R_m \gg R_{pip}.$$

5. Depending on the question to be answered **different configurations of patch-clamp** method are possible:
    (a) **Cell-attached**: Cell responses under physiological conditions.
    (b) **Inside-out**: Regulation of intracellular messaging.
    (c) **Outside-out**: Regulation by extracellular drugs.
    (d) **Whole-cell**: Multi-channel and carrier transport analysis.
6. **Single-channel data** provide information about **single-channel conductance** and **open-close kinetics** and their modulations by various parameters (e.g. membrane potential or interaction with drugs).
7. To screen (e.g. for the effects of drugs) a large number of cells **automated voltage-clamp techniques** have been developed for the patch-clamp method.

Exercises

1. Prepare a table of the different voltage-clamp versions with their characteristic applications.
2. What can be extracted from current fluctuations, and what are the respective assumptions?
3. What can be extracted from transient charged movements?
4. What are the advantages of automated electrophysiology compared to manual methods?
5. Why is pharmaceutical industry interested in automated electrophysiology?

6. What is the crucial difference between electrophysiology and optical or binding assays when characterising ion channels or electrogenic transporters?
7. Why is TEVC on oocytes still a useful method?
8. What is the basis of most automated patch-clamp systems?

References

Adrian, R. H., Chandler, W. K., & Hodgkin, A. L. (1970). Voltage clamp experiments in striated muscle fibres. *The Journal of Physiology, 208*, 607–644.

Allen, T. G. J. (1997). The 'sniffer-patch' technique for detection of neurotransmitter release. *Trends in Neurosciences, 20*, 192–197.

Cole, K. S. (1949). Dynamic electrical characteristics of the squid axon membrane. *Archives des Sciences Physiologiques, 3*, 253–258.

Fertig, N., Blick, R. H., & Behrends, J. C. (2002). Whole cell patch clamp recording performed on a planar glass chip. *Biophysical Journal, 82*, 3056–3062.

Fuhrmann, G. F., Schwarz, W., Kersten, R., & Sdun, H. (1985). Effects of vanadate, menadione, and menadione analogs on the Ca^{2+}-activated K^+ channels in human red blood cells: Possible relations to membrane-bound oxidoreductase activity. *Biochimica et Biophysica Acta, 820*, 223–234.

Hille, B. (1992). *Ionic channels of excitable membranes* (2nd ed.). Sinauer Associates.

Hille, B. (2001). *Ionic channels of excitable membranes* (3rd ed.). Sinauer Associates.

Hodgkin, A. L., Huxley, A. F., & Katz, B. (1952). Measurement of current-voltage relations in the membrane of the giant axon of Loligo. *Journal of Physiology (London), 116*, 424–448.

Kostuk, P. G., & Krishtal, O. A. (1984). Intracellular perfusion of excitable cells. In *IBRO handbook series: Methods in the neurosciences* (Vol. 5). Wiley.

Muller-Chretien, E. (2014). Outside-out "Sniffer-Patch" clamp technique for in situ measures of neurotransmitter release. In M. Martina & S. Taverna (Eds.), *Patch-clamp methods and protocols* (pp. 195–204). Springer.

Neher, E. (1991). Ion channels for communication between and within cells. In *Nobel lectures 1991–1995*.

Neher, E., & Sakmann, B. (1976). Single-channel currents recorded from membrane of denervated frog muscle fibres. *Nature, 260*, 799–802.

Nonner, W. (1969). A new voltage clamp method for Ranvier nodes. *Pflügers Archiv, 309*, 176–192.

Rettinger, J., & Schwarz, W. (1994). Ion-selective channels in K562 cells: A patch-clamp analysis. *Journal of Basic and Clinical Physiology and Pharmacology, 5*, 1–18.

Rettinger, J., Vasilets, L. A., Elsner, S., & Schwarz, W. (1994). Analysing the Na^+/K^+-pump in outside-out giant membrane patches of Xenopus oocytes. In E. Bamberg & W. Schoner (Eds.), *The sodium pump* (pp. 553–556). Steinkopff Verlag.

Sakmann, B. (1991). Elementary steps in synaptic transmission revealed by currents through single ion channels. In *Nobel lectures 1991–1995*.

Sakmann, B., & Neher, E. (1995). *Single-channel recording* (2nd ed.). Plenum Press.

Schwarz, W. (1983). Sodium and potassium channels in myelinated nerve fibers. *Experientia, 39*, 935–941.

Schwarz, W., Grygorczyk, R., & Hof, D. (1989). Recording single-channel currents from human red-cells. *Methods in Enzymology, 173*, 112–121.

Sigworth, F. J., & Sine, S. M. (1987). Data transformations for improved display and fitting of single-channel dwell time histograms. *Biophysical Journal, 52*, 1047–1054.

Stämpfli, R. (1954). A new method for measuring membrane potentials with external electrodes. *Experientia, 10*, 508–509.

Taglialatela, M., Toro, L., & Stefani, E. (1992). Novel voltage clamp to record small, fast currents from ion channels expressed in Xenopus oocytes. *Biophysical Journal, 61*, 78–82.

Taylor, R. E. (1963). Cable theory. In W. L. Nastuk (Ed.), *Physical techniques in biological research* (pp. 219–262). Academic Press.

Wu, Y., Huang, M., Xia, Y., & Ding, G. H. (2019). Real-time analysis of ATP concentration in acupoints during acupuncture: A new technique combining microdialysis with patch clamp. *Journal of Biological Engineering, 13*, 93.

Ion-Selective Channels

5

Contents

Abstract

The major conductance pathways in cell membranes, the ion-selective channels, shall be described in this chapter with respect to their specific characteristics. The selectivity of an ion channel for a specific ion is explained, and consequences of discrete ion movement through a pore are discussed and illustrated by examples.

Keywords

Ion selectivity · Single-file movement

Ion-selective conductances of a cell membrane are mediated by electrogenic carriers or channel proteins. We will now discuss in more detail the channels, which are the dominating ion pathways that govern the membrane potential. Carriers will be dealt with in Chap. 7. The differences between different channel types are based on characteristics we have discussed before:

© The Author(s), under exclusive license to Springer Nature Switzerland AG 2022

J. Rettinger et al., *Electrophysiology*,

https://doi.org/10.1007/978-3-030-86482-8_5

1. Selectivity for ion species
2. Single-channel conductance
3. Gating mechanisms.

5.1 General Characteristics of Ion Channels

In this section we will gather properties that characterise the ion channels.

5.1.1 Selectivity of Ion Channels

First, we want to ask the question: What makes an ion channel permeable for one ion species but not for another one?

When we discussed the GHK equation for membrane potential (Sect. 2.4), we have seen that the changes in membrane potential are based on the changes in ion permeabilities. At the resting potential the permeability P_K for K^+ is high, during an action potential the permeability P_{Na} for Na^+ temporarily rises exceeding finally P_K. Of course, the changes in potential are only possible if activity gradients for the respective ion species exist.

With this knowledge we have a tool in our hands that allows to determine permeability ratios using the GHK equation. For example, we can vary the concentrations in the extracellular medium and measure the corresponding membrane potentials. It would be helpful to know the intracellular ion concentrations or activities. This can be achieved, e.g. with the help of the ion-selective microelectrodes (see Sect. 3.4.3). But we have learned that it is also possible to manipulate the intracellular ion composition (*perfused axons, cut fibres, excised membrane patches*). Under such conditions determination of permeability ratios can become very simple by using *bi-ionic* conditions (see Sect. 2.4); then the GHK equation takes the form:

$$E_{rev} = \frac{RT}{zF} \ln \left(\frac{P_A[A_o]}{P_B[B_i]} \right).$$

For Na^+ and K^+-selective channels in excitable membranes of nerve the relative permeabilities were calculated to (see e.g. also Hille, 2001):

Na$^+$ channel: **Na$^+$** (1): **Li$^+$** (0.93): **Guanidinium** (0.12): **K$^+$** (0.09): **TMA$^+$** (<0.005)

K$^+$ channel: **K$^+$** (1): **Tl$^+$** (2.3): **Rb$^+$** (0.91): **Na$^+$** (0.01): **TEA$^+$** (=0)

If we compare the extreme ends of the selectivity sequences, we realise that the channels are well permeable for small cations, large ones can hardly permeate. This led to the idea that channels have some kind of sieve filtering the ions according to their size (Fig. 5.1). In Hille's hypothesis the selectivity filter is formed by a ring of carbonyl oxygen of 3×5 Å for the Na^+ channel allowing cations up to this size to cross the membrane.

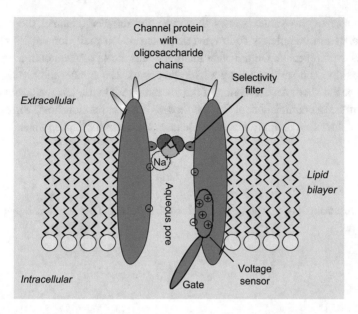

Fig. 5.1 The Hille idea of a selectivity filter (as example the Na$^+$ channel, the Na$^+$ with one hydrogen ion is shown) compare also Hille (2001, Fig. 3.14). A ring of carbonyl oxygen has been postulated to form the filter. The figure cartoons the situation for a Na$^+$ channel with an internally oriented voltage-sensitive gate

Table 5.1 Permeability ratios for Na+ and K+ channels (determined from measurements of reversal potentials, see also Hille, 2001)

Na$^+$ channels		K$^+$ channels	
P_x/P_{Na}	Ion$_x$	P_x/P_K	Ion$_x$
1	Na$^+$	<0.018	Li$^+$
0.94	Hydroxylammonium	0.01	Na$^+$
0.93	Li$^+$	1	K$^+$
0.59	Hydrazinium	2.3	Tl$^+$
0.16	NH$_4^+$	0.91	Rb$^+$
0.13	Guanidinium	0.13	NH$_4^+$
0.086	K$^+$	0.029	Hydrazinium
<0.013	Cs$^+$	<0.077	Cs$^+$
<0.012	Rb$^+$	<0.013	Guanidinium

This idea is based on the finding that ions exceeding a certain size cannot permeate. What about smaller cations? The alkali ions have crystal radii (in 10^{-8}-cm) of:

$$\textbf{Li}^+ \ (0.6) - \textbf{Na}^+ \ (0.95) - \textbf{K}^+ \ (1.33) - \textbf{Rb}^+ \ (1.48) - \textbf{Cs}^+ \ (1.69).$$

This sequence could roughly account for the selectivity of a Na$^+$ channel. Table 5.1 shows once more measured selectivity ratios for Na$^+$ and K$^+$ channels of frog nerve:

Comparison of the values shows that even for the Na^+ channel the permeability sequence does not completely follow the sequence of ion radii; for the K^+ channel the sequence is even reversed. On the other hand, the ions are of course not in a crystal but in aqueous solution and surrounded by a hydration shell. The smaller the crystal radius the larger will be the shell. An estimation of the ion radii with hydration shell can be obtained from the diffusion coefficient applying the Stokes–Einstein relationship (keep in mind that it may be problematic to apply macroscopic laws to microscopic phenomena):

$$D = \frac{kT}{6\pi\eta r}$$

with η representing the viscosity (for water 0.01 Poise). This relationship is based on 1. Fick's law:

$$\frac{dn}{dt} = DA\frac{dc}{dx}.$$

The radii will be
for Na^+: $r = 2.4\text{–}3.3\ 10^{-8}$ cm,
and for K^+: $r = 1.6\text{–}2.2\ 10^{-8}$ cm.
The result is opposite to that for the crystal radii. This offers the possibility that selectivity for Na^+ channels is more governed by the crystal radii, and for K^+ channels more by the radii of the hydrated ions. According to the Hille idea, the ions have to pass the filter where they interact with negatively charged sites and partially lose their hydration shell. The relative affinities of a binding site for two different ions **a** and **b** are given by the differences of free enthalpies:

$$\Delta G = (\Delta G_{Ba} - \Delta G_{Bb}) - (\Delta G_{H_2Oa} - \Delta G_{H_2Ob})$$

with ΔG_B: change in Gibb's energy on binding
ΔG_{H_2O}: change in hydration energy.
If $\Delta G > 0$, we would have for the corresponding ion permeabilities $P_a > P_b$, and vice versa. If binding is based on pure Coulomb interaction, we have for the binding energy:

$$U_B = \frac{z_B z_i e^2}{4\pi\varepsilon\varepsilon_0(r_B + r_i)}$$

with r_B and r_i representing the effective radii for binding site and ion, respectively.
Let us consider two extreme cases

1. r_B is small (binding site with high electric field strength)

Table 5.2 Eisenman sequences of ion selectivity (see Eisenman, 1962)

Group	*Weak field strength binding sites*								
I	Cs^+	>	Rb^+	>	K^+	>	Na^+	>	Li^+
II	Rb^+	>	Cs^+	>	K^+	>	Na^+	>	Li^+
III	Rb^+	>	K^+	>	Cs^+	>	Na^+	>	Li^+
IV	K^+	>	Rb^+	>	Cs^+	>	Na^+	>	Li^+
V	*K^+*	>	*Rb^+*	>	*Na^+*	>	*Cs^+*	>	*Li^+*
VI	K^+	>	Na^+	>	Rb^+	>	Cs^+	>	Li^+
VII	Na^+	>	K^+	>	Rb^+	>	Cs^+	>	Li^+
VIII	Na^+	>	K^+	>	Rb^+	>	Li^+	>	Cs^+
IX	Na^+	>	K^+	>	Li^+	>	Rb^+	>	Cs^+
X	*Na^+*	>	*Li^+*	>	*K^+*	>	*Rb^+*	>	*Cs^+*
XI	Li^+	>	Na^+	>	K^+	>	Rb^+	>	Cs^+
	High-field strength binding sites								

$\Delta G_B >> \Delta G_{H2O}$, hence the selectivity is governed by $\Delta G_{Ba} - \Delta G_{Bb}$, and we will have a sequence of $Li^+ > Na^+ > K^+ > Rb^+ > Cs^+$ governed by crystal radii

2. r_B is large (binding site with low electric field strength)

$\Delta G_B \ll \Delta G_{H2O}$, hence the selectivity is governed by $\Delta G_{H2Oa} - \Delta G_{H2Ob}$, and we will have a sequence of $Li^+ < Na^+ < K^+ < Rb^+ < Cs^+$ governed by radii of hydrated ions.

For intermediate field strengths different combinations are possible. Eisenman (1962) found that from the 120 possible combinations only 11 could be observed (see Table 5.2), in his experiments with ion-selective glasses. According to his nomenclature the Na^+ channel belongs to group **X** and has a high-field strength site, the K^+ channels belong to group **V** with a selectivity filter of intermediate field strength site.

5.1.2 Discrete Movement of Ions through Pores

In our previous description of ion movement across the membrane we considered principles of free diffusion. In the section on selectivity, we have seen that specific interaction of a selectively permeant ion with a binding site has to be considered. Additional interaction during the passage of an ion through a pore has been introduced to explain deviations from independence and from the GHK equations (Hodgkin and Keynes, 1955). Examples of deviations from predictions of independence (see Hille and Schwarz, 1978) include those from Ussing flux ratio, dependence of conductance on concentration, or from anomalous mole-fraction behaviour (see also Sects. 5.2.2, 5.2.3, and 5.2.4).

Discrete movement of ions interacting with a sequence of binding sites can be described by passage through an energy profile using rate coefficients k for the jumps (see Fig. 5.2) according to Eyring rate theory (Glasstone et al., 1941):

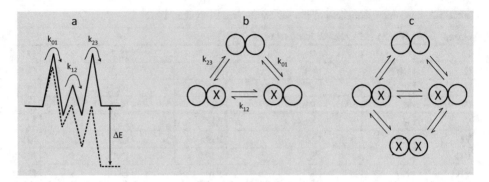

Fig. 5.2 Energy profile for a two-site pore (**a**) and state diagram for a one-ion (**b**) and a two-ion (**c**) pore. The Xs refer to occupied sites

$$k = q\frac{RT}{h}e^{-\frac{\Delta G}{RT}}, \text{with } \Delta G = \Delta G_B + zEF$$

with frequency factor $q\frac{RT}{h}$, where q is the transmission coefficient (usually assumed to be 1) and h the Planck's constant. ΔG represents the barrier height, E a superimposed electrical potential difference.

In the simplest case, only a single ion can occupy a pore at a time, but also multiple occupancy could be possible as illustrated in Fig. 5.2. Different predictions for independent ion movement, one-ion and multi-ion pores are obtained for the dependence of conductance on ion concentration as illustrated in Fig. 5.3a. For a single-ion pore the exit rate of an ion from the channel becomes rate-limiting at high concentration, and hence the conductance saturates. For a two-ion pore a maximum can be obtained since at high concentration double-occupied states become populated (compare Fig. 5.5) but, on the other hand, empty sites inside the pore are necessary for ion transit.

To handle the transport based on ion-hopping through a narrow pore, a mathematical tool can be applied that has been developed long time ago to describe electrical networks in terms of Kirchhoff's laws (Kirchhoff, 1847) and later independently by King and Altman to describe enzyme reactions (King & Altman, 1956). A simple outline of this so-called Graph Theory (for details see Rettinger et al. 2018).

To explain deviations from Ussing flux ratio, multiple occupancy and movement of ions in single-file fashion through a pore has been introduced, which leads to a modified flux ratio equation:

$$\left|\frac{I_{eff}}{I_{in}}\right| = \left(\frac{c_i}{c_o}e^{zEF/RT}\right)^n$$

with $n > 1$. Depending on the number of sites m and the degree of occupancy, n can have values of $1 \leq n \leq$ m.

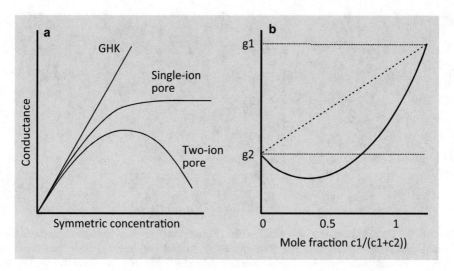

Fig. 5.3 Dependence of conductance on concentration (**a**) and anomalous mole-fraction behaviour (**b**)

Another characteristic of several channel types that cannot be explained by independent ion movement is voltage-dependent inhibition by a blocking ion with an effective charge larger than the valency of the blocking ion (see Fig. 5.6) and the anomalous mole-fraction behaviour (see Fig. 5.3b) . In the latter case, stepwise replacement of an ion species c2 by an ion species c1 with higher permeability can lead to a reduced conductance (see also Fig. 5.7).

5.2   Specific Ion Channels

5.2.1   The Na$^+$ Channel (A Single-Ion Pore)

To simulate several characteristics of the Na$^+$-selective channel of excitable membranes, Hille has applied a three-site, one-ion pore model (see Fig. 5.4).

With this model the concentration dependence, which deviates from independence, could be described as well as ion selectivity by modulation of the barrier C. To give the selectivity ratios for Li$^+$ and K$^+$ as listed in Table 5.1, the ΔG values were (in RT)

$$\Delta G_{Li} = \Delta G_{Na} + 0.1$$

$$\Delta G_K = \Delta G_{Na} + 2.7$$

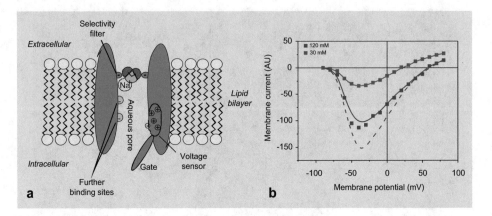

Fig. 5.4 (a) Three-binding-site model for the Na^+ channel and (b) current–voltage dependencies for 30 mM (red symbols) and 120 mM (blue symbols) external Na^+ activity. Blue lines represent predicted dependencies based on GHK (dashed) and on single-file movement (solid) based on the dependency at 30 mM Na^+ (for real data compare Hille, 1992)

illustrating that small changes in barrier height can lead to pronounced changes in selectivity (Hille, 1992). The binding site B represents the high-field strength site (possibly formed by $-COO^-$ group), where part of the hydration shell can be removed so that the Na^+ can pass with only 3 H_2O left. This model also illustrates that the restricted passage of the ion occurs within a few Å only.

An interesting phenomenon found for Na^+ channels is the so-called permeability paradox for H^+ compared to Na^+. From the GHK equation permeability ratio can be determined to:

$$P_H/P_{Na} = 250$$

or in terms of the energy barrier C ($P \propto \exp(-C/RT)$):

$$C_H/C_{Na} = \ln(250).$$

Accordingly, an increase in $H^+{}_o$ should increase the inward current, but this is not observed. In contrast, reduction of external pH leads to inhibition of current with a $pK_a = 5.4$, which corresponds to $K_H = 4$ μM, and in the absence of Na^+ only a tiny current can be detected. For the dependence of current on Na^+ concentration a K_{Na} value of 400 mM was obtained. Assuming that H^+ as well as Na^+ temporarily bind to the site B, the site B_H should be deeper than B_{Na}:

$$B_H/B_{Na} = \ln(10^5/1).$$

The difference in barrier height is, therefore,

$$\Delta G = RT(\ln 10^5 - \ln 250) = RT \ln 400 \text{ and } I_H/I_{Na} = 1/400.$$

In other words, the paradox can be explained by assuming that binding of a proton blocks Na^+ permeation because the proton leaves the binding site 400 times slower than the Na^+.

5.2.2   The K⁺ Channel (A Multi-Ion Pore)

The K^+-selective channels exhibit several features of long multi-ion pores (Hille & Schwarz, 1978):

1. Flux ratio exponents have been determined in the range of $n = 2$–2.5. This suggests that K^+ channels should have at least three sites (see Sect. 5.1.2).
2. In principle maxima in the conductance versus concentration curve should be expected (Hille & Schwarz, 1978), depending on the number of binding sites m the number of maxima can be up to m-1 (compare Fig. 5.5b). The concentration range that can be covered experimentally is usually not large enough, and often only saturation could be observed. Lu and MacKinnon (1994) could demonstrate for an inward-rectifying K^+ channel a maximum of single-channel conductance at about 300 mM (Fig. 5.5a). Because the conductance is independent of membrane potential, the inward rectification cannot be attributed to reduced outward conductance (compare Fig. 5.5).
3. The voltage-dependence of block by a blocking ion X can be described by Boltzmann distribution for the relative ratio of unblocked (1-B) to blocked channels blocked (B). For a one-to-one inhibition we have

$$\frac{1-B}{B} = e^{\frac{-zF(E-EB)}{RT}}$$

with EB representing the potential for 50% inhibition.

As discussed above, for a single-ion pore an effective valency z not larger than the valency of the blocking ion should be expected. In fact, for inhibition of K^+ current in starfish egg by Cs^+ (see Fig. 5.6a and compare Hagiwara et al., 1976) an effective valency of 1.8 was determined. For an m-site multi-ion pore and a blocking ion of valency 1 values of $z \leq$ m are possible. A value of $z = 4$ was used to describe the voltage-dependent inhibition (Fig. 5.6b).

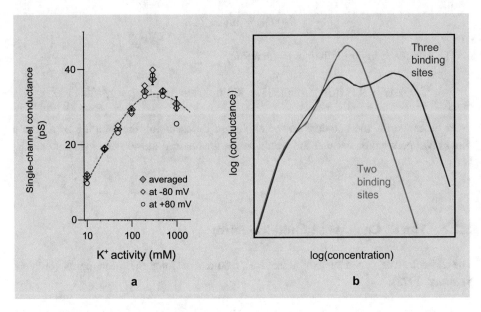

Fig. 5.5 (**a, b**) Conductance versus extracellular K+ activity. Data represent values at −80 and +80 mV and averaged for this potential range (based on Lu & MacKinnon, 1994; Hille & Schwarz, 1978)

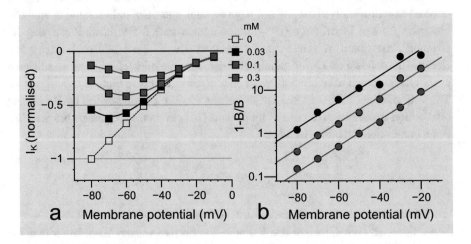

Fig. 5.6 Inhibition of inward K^+ current by external Cs^+ (**a**) Current–voltage dependencies at different concentrations of Cs^+ in the external solution (based on data from Hagiwara et al., 1976). (**b**) Voltage-dependence of current inhibition. Solid lines represent fits of Boltzmann distribution with an effective valency of 4

4. K^+ channels show anomalous mole-fraction behaviour for Tl^+ compared to K^+ with higher conductance for Tl^+ in solutions that contain only K^+ or Tl^+, respectively (compare Fig. 5.7). This can be explained by stronger binding of K^+ than of Tl^+ (Fig. 5.7b) and pronounced electrostatic repulsion if two ions want to enter the pore simultaneously (see Hille & Schwarz, 1978).

5. A typical phenomenon of inward-rectifying K^+ channels is the "crossing over" of current–voltage curves if the external K^+ concentration is raised (Fig. 5.8). Inward rectification can be explained by an internal blocking ion; theoretically this could be a cytoplasmic ion or a charged, flexible cytoplasmic domain of the channel protein. Elevation of extracellular K^+ could counteract the entering of the blocking particle, and hence increase the conductance.

5.2.3 The Ca^{2+} Channel (A Multi-Ion Pore)

Also the Ca^{2+}-selective channels show multi-ion characteristics. This includes

1. anomalous mole-fraction behaviour between Ca^{2+} and Ba^{2+} (Ba^{2+} having the higher permeability when measured in the presence of only one of the ion species).

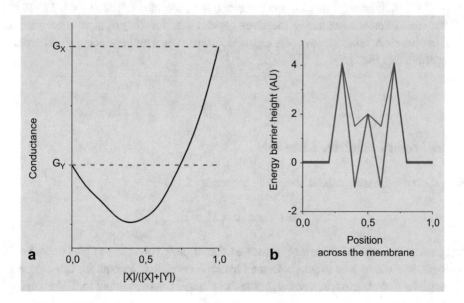

Fig. 5.7 Anomalous mole-fraction behaviour for an ion channel with two binding site and with conductance for ion G_X (blue) and ion G_Y (red) (**a**) and the corresponding energy profile for the binding sites (**b**)

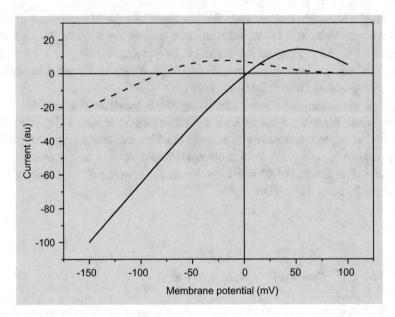

Fig. 5.8 Crossing over of inward-rectifying current–voltage curves. Broken line in normal external K^+, solid line at high external K^+ ($K^+_o = K^+_I$)

2. the finding that the channel is permeable for Na^+, and low concentrations of Ca^{2+} < 0.1 mM (if no other divalent ion is present) have blocking effects ($K_I = 0.5\ \mu M$), but the conductance dramatically increases beyond that for Na^+ at high concentration. Electrostatic repulsion leads to the high conductance when multiple occupancy becomes possible (see Fig. 5.9).

5.2.4   Anion-Selective Channels

Cl^--selective channels exhibit selectivity sequence of

$$Br^- \approx J^- > Cl^- > F^-$$

suggesting in terms of Eisenman sequences that a positively charged site of low-field strength, that allows ions to pass with their hydration shell, can account for this sequence. Anomalous mole-fraction behaviour for the Cl^- channel has been found for Cl^- compared to SCN^-.

Varying KCl concentration yielded that the E_{rev} does not follow the Nernst potential for chloride E_{Cl}, but a permeability ratio of $P_K/P_{Cl} \approx 0.2$ can be estimated from application of

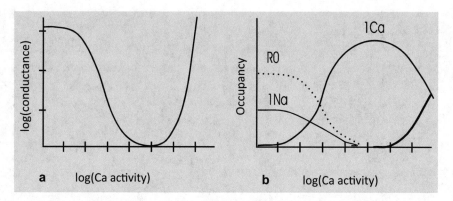

Fig. 5.9 The Ca^{2+} channel as a multi-ion pore. Dependence of conductance on Ca^{2+} activity (**a**) explained by changes in probability of occupancy (**b**) (based on Almers & Mccleske, 1984, Fig. 11, with kind permission from John Wiley and Sons, 1984)

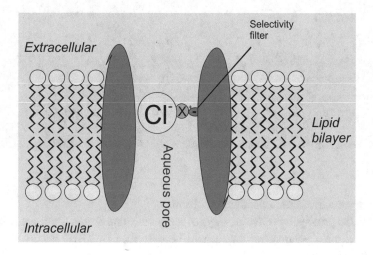

Fig. 5.10 Model of dipole site in a Cl^{-}-permeable channel where X^{+} represents a cation with low probability of passage (compare also Franciolini & Nonner, 1987)

GHK equation. On the other hand, in the absence of permeant anions (Cl^{-} replaced by e.g. SO_4^{2-}) no cation permeability could be detected. This has been explained by assuming that a cation binds to a negatively charged site inside the pore (Fig. 5.10) leading to a low-field dipole site that allows the Cl^{-} to pass. Only occasionally also the cation may pass together with the anion. In absence of Cl^{-} the cation cannot leave the channel and no permeability is detected.

Take-Home Messages

1. **Selectivity sequences** can be described according to ion radius and "**high-**" and "**low-field**" binding sites, which allow removal of hydration shell.
2. Examples for **Discrepancies** between phenomena and prediction described by continuous **(GHK)** can be attributed to discrete **(single-file)** ion movements, e.g.:
   (a) Concentration dependence of conductance.
   (b) Anomalous mole-fraction behaviour.
   (c) Deviation from Ussing flux equation.
   (d) Selectivity paradox.
   (e) Effective charge of a blocking ion large than the real value.
   (f) Cross-over effect.
3. **Na^+ channel** can be described as a **single-ion pore** (to describe, e.g. concentration dependence of conductance, selectivity paradox (Na^+/H^+)).

   K^+ channel can be described as a **multi-ion pore** (to describe, e.g. anomalous-mole-fraction behaviour, voltage-dependent block by Cs^+ with $z_{eff} > 1$, deviation from Ussing flux ratio, cross-over effect, anomalous mole-fraction behaviour (K^+/Tl^+).

   Ca^{2+} channel can be described as a **multi-ion pore** (to describe, e.g. anomalous-mole-fraction behaviour (Ca^{2+}/Na^+).

   Cl^- channel can be described as a **multi-ion pore** (to describe, e.g. anomalous mole-fraction behaviour (Cl^-/SCN^-), permeability paradox (Cl^-/K^+).
4. If **GHK equation** is used to calculate e.g. **permeability ratios**, the result should be **interpreted with reservation**.

Exercises

1. How can we explain ion selectivity of channels?
2. Describe the principally different ways of ion permeation across a cell membrane.
3. List and explain phenomena that cannot be described by GHK but by discrete ion movements.
4. Describe and explain characteristics of Na^+-selective channels.
5. Describe and explain characteristics of K^+-selective channels.
6. Describe and explain characteristics of Ca^{2+}-selective channels.
7. Describe and explain characteristics of Cl^--selective channels.
8. How can the selectivity of a channel be determined experimentally?

References

Almers, W., & Mccleske, E. W. (1984). Non-selective conductance in calcium channels of frog muscle: Calcium selectivity in a single-file pore. *The Journal of Physiology, 353*, 585–608.

Eisenman, G. (1962). Cation selective glass electrodes and their mode of operation. *Biophysical Journal, 2*, 259–323.

Franciolini, F., & Nonner, W. (1987). Anion and cation permeability of a chloride channel in rat hippocampal neurons. *The Journal of General Physiology, 90*, 453–478.

Glasstone, S. K., Laidler, J., & Eyring, H. (1941). The theory of rate processes McGraw-Hill Book Company.

Hagiwara, S., Miyazaki, S., & Rosenthal, N. P. (1976). Potassium current and the effect of cesium on this current during anomalous rectification of the egg cell membrane of a starfish. *The Journal of General Physiology, 67*, 621–638.

Hille, B. (1992). *Ionic channels of excitable membranes*. Sinauer Associates.

Hille, B. (2001). *Ionic channels of excitable membranes*. Sinauer Associates.

Hille, B., & Schwarz, W. (1978). Potassium channels as multi-ion single-file pores. *The Journal of General Physiology, 72*, 409–442.

Hodgkin, A. L., & Keynes, R. D. (1955). The potassium permeability of a giant nerve fibre. *The Journal of Physiology, 128*, 61–88.

King, E. L., & Altman, C. (1956). Schematic method of deriving the rate laws for enzyme-catalyzed reactions. *The Journal of Physical Chemistry, 60*, 1375–1378.

Kirchhoff, G. (1847). Ueber die Auflösung der Gleichungen, auf welche man bei der Untersuchung der linearen Vertheilung galvanischer Ströme geführt wird. *Poggendorfs Annalen der Physik und Chemie, 72*, 497–508.

Lu, Z., & MacKinnon, R. (1994). A conductance maximum observed in an inward-rectifier potassium channel. *The Journal of General Physiology, 104*, 477–486.

Rettinger, J., Schwarz, S., & Schwarz, W. (2018). *Elekrophysiologie: Grundlagen, Methoden, Anwendungen*. Springer.

Theory of Excitability

6

Contents

Abstract

Essential for membrane excitability is the action potential. The generation and propagation is based on the Hodgkin–Huxley description, which shall be presented in this chapter. Time- and potential-dependent changes of Na^+ and K^+ permeability account for generation, and spread of the action potential, and the synaptic transmission shall also be presented in this chapter.

© The Author(s), under exclusive license to Springer Nature Switzerland AG 2022   101
J. Rettinger et al., *Electrophysiology*,
https://doi.org/10.1007/978-3-030-86482-8_6

Keywords

Hodgkin–Huxley description · Action potential · Electrotonic potential · Continuous
spread · Saltatory spread · Synaptic transmission · Surface potential

In the previous chapter we have discussed in detail the selectivity of ion channels and ion
permeation processes. Another characteristic of pores is their gating, the kinetics of their
opening and closing. We have already mentioned this property when we discussed
characterisation of single channels by patch-clamp technique (Sect. 4.4) and analysis of
current fluctuations (Sect. 4.2) and gating currents (Sect. 4.3). The patch-clamp technique
particularly introduced a new "microscopic" dimension for the understanding of electrophysiological phenomena. In the following we will consider "macroscopic" phenomena that,
nevertheless, formed a milestone in electrophysiology. Macroscopic in this context means
that current signals are generated by a huge number of simultaneously active ion channels.

6.1 The Hodgkin–Huxley Description of Excitation

6.1.1 Experimental Basics

After the discovery that excitability of nerve and muscle cells is based on the spread of
action potentials along the cell fibre (nerve-muscle preparations), Bernstein (1902, 1912)
hypothesised that the resting potential is based on a selective permeability of the cell
membrane for K^+ ions. During excitation a reversible breakdown of selectivity was
postulated. This should lead to a change in membrane potential (action potential) to
about -15 mV ($=$ Donnan potential, Sect. 2.3).

Detailed analysis of the action potential was first done by Hodgkin, Huxley, and Katz at
the giant nerve fibre of the squid (see Fig. 6.1). They found that reduction of extracellular
Na^+ led to reduction in the height of action potential (see Table 6.1). These findings led to
the hypothesis that the Na^+ permeability increases during excitation becoming much larger
than the permeability for K^+. This was the state of knowledge in 1949. The demonstration
and quantitative description of an action potential by ion-selective currents was achieved
by Hodgkin and Huxley in a series of papers published in 1952 in the Journal of Physiology
(see Hodgkin & Huxley, 1952). The basis for their work was the voltage-clamp technique
and the separation of the different current components.

Figure 6.2 illustrates schematically the current signals in response to a hyperpolarising
and a depolarising voltage-clamp pulse. The hyperpolarisation leads to transient capacitive
and a steady state unspecific, the so-called leak current of ohmic type. Depolarisation also
leads to a transient capacitive current followed (in the millisecond range) by a transient
inward current that finally reverses to a steady-state outward current. If external Na^+ is
replaced by the impermeable cation choline, the transient inward current is completely

Fig. 6.1 Schematic drawing of the effect of extracellular Na^+ activity on action potential amplitude. A 50% reduction of Na^+ leads to a decrease of peak of action potential of $\Delta E_{AP} = 21$ mV

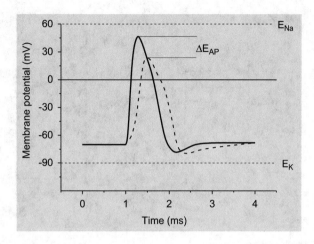

Table 6.1 Effect of reducing extracellular Na^+ activity on action potential (values adapted from Hodgkin & Katz, 1949)

	measured	Predicted by Bernstein	Predicted by Hodgkin, Huxley, Katz
Action potential	+30 − +50 mV	$E_{Donnan} = -15$ mV	$E_{Na} = +53$ mV
$Na_o^+ \rightarrow \frac{1}{2}$ Na_o^+	$\Delta E_{AP} = 21$ mV		$\Delta E_{AP} = \Delta E_{Na} = 17$ mV

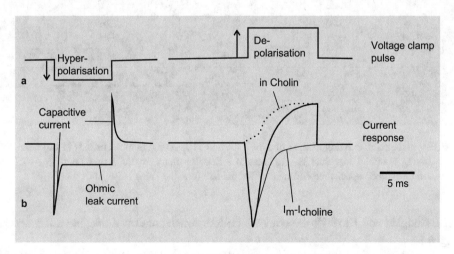

Fig. 6.2 (**a, b**) Schematic current signals (**b**) in response to voltage-clamp pulses (**a**). The transient inward-directed current is blocked if Na^+ is replaced by choline.

blocked, leaving only a delayed outward current. This suggests that the transient current is mediated by Na^+.

An important step was the separation of this time-dependent current into two components one carried by inward-moving Na^+ ions, the other by outward-moving K^+

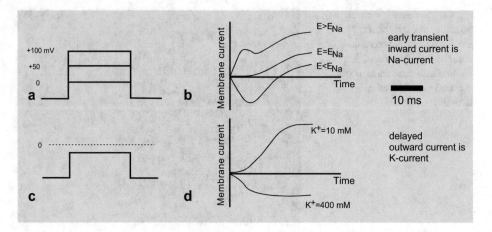

Fig. 6.3 Demonstration of fast activating and slowly inactivating Na$^+$ currents in response to different depolarising voltage-clamp pulses (**a, b**), and of slowly activating K$^+$ currents in response to a depolarising pulse to a still negative potential (**c, d**). Left: depolarising voltage pulses, right: respective current responses

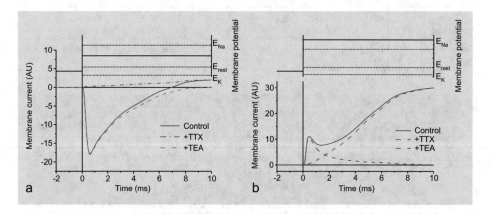

Fig. 6.4 Schematic drawing of the effect of TTX and TEA on membrane currents in response to a depolarising potential less than E$_{Na}$ (**a**) and to a depolarising potential larger than E$_{Na}$ (**b**). The unspecific leak and capacitance currents are subtracted. (compare also Hille, 2001)

ions. Hodgkin and Huxley's experiments (1952), which confirmed this, are illustrated in Fig. 6.3.

The early transient current reverses at the Nernst potential for Na$^+$ (Fig. 6.3 a,b) and this observation proved that the current is carried by Na$^+$. This is also confirmed by experiments with the highly specific Na$^+$ channel inhibitor tetrodotoxin (TTX, see Fig. 6.4). The delayed outward current can also easily be analysed by replacement of Na$^+$ by choline (Fig. 6.3c, d). The remaining current reverses at high external K$^+$ while Cl$^-$ is kept constant and can be inhibited by the K$^+$-channel blocker tetraethylammonium (TEA. Figure 6.4).

This demonstrates that the delayed current is carried by K^+ and not by Cl^-. Later on, this could be demonstrated directly by comparison of radioactive tracer fluxes with the currents (Hodgkin & Keynes, 1955). Blocking the Na^+ as well as the K^+ conductances leaves in addition to the capacitive current the unspecific, time- and voltage-independent leak conductance.

Altogether, the results of the experiments, therefore, allowed to describe the total membrane current by the sum of several independent components:

$$I_m(t, E) = C\frac{dE}{dt} + I_{Leak}(E) + I_{Na}(t, E) + I_K(t, E)$$

These were the starting point for Hodgkin–Huxley description of an action potential. For the development of the model two requirements were essential:

– The application of the voltage-clamp technique to the squid axon, and
– The separation of the ionic currents.

This made it possible to perform a detailed analysis of all the kinetic properties of ion-selective conductances that finally allowed Hodgkin and Huxley to describe the time course of an action potential based on these conductances.

6.1.2 The Hodgkin–Huxley (HH) Description of Excitability

The Hodgkin–Huxley model assumes that an ion-selective channel can either be open or closed, an assumption that was also made later on to analyse current fluctuations, but which was proven only 30 years later by the application of the patch-clamp technique by Neher and Sakmann (1976).

6.1.2.1 The Hypothetical Channel
The membrane conductance g originating from N channels of conductance γ is given by

$$g = N \gamma\ p(t, E)$$

with p being the probability of a channel to be open. Assuming a gating particle that changes orientation with the changes in the electrical field and that has an open- and closed-state position (compare Fig. 6.5a), p can be described by

$$\frac{dp}{dt} = \alpha(1 - p) - \beta p \quad \text{with the solution} \quad p(t) = p_\infty - (p_\infty - p_0)e^{-t/\tau}$$

$$\text{where} \quad p_\infty = \frac{\alpha}{\alpha + \beta} \quad \text{and} \quad \tau = \frac{1}{\alpha + \beta}.$$

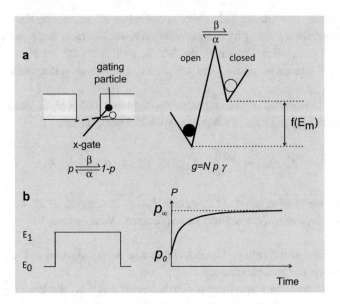

Fig. 6.5 Illustration of a hypothetical channel with channel-open probability p. (**a**) cartoon of channel and energy profile. (**b**) Time course of p in response to a voltage step

Figure 6.5b illustrates the change of open probability with time in response to a sudden change in potential.

6.1.2.2 The K^+ Channel

The activation of the K^+ channel does not follow a pure exponential time course as expected from the simple hypothetical channel. Therefore, Hodgkin and Huxley assumed the existence of several gating particles that have to be simultaneously in the open-state position n to make the channel conducting. To describe the time course 4 such independently moving particles were necessary (compare Fig. 6.6): The probability of a K^+ channel for being open will then be n^4, and

$$g_K = N \ \gamma_K \ n^4 = g*_K \ n^4$$

with maximum conductance g^*_K and n having time and voltage dependence (Fig. 6.7) qualitatively similar to that described above for p.

At the resting potential the n-gates are predominantly closed ($n \approx 0$). In response to a depolarising pulse n increases to a new potential-dependent steady state n_∞. The

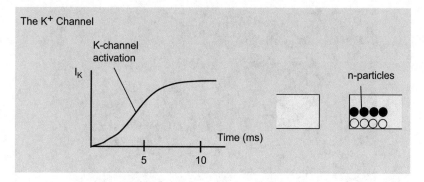

Fig. 6.6 HH model for the K$^+$ channel with 4 gating particles n that can be in closed or open position

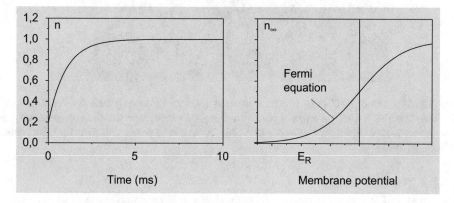

Fig. 6.7 Time and voltage dependence of n

distribution of the probability of an n particle being in the open-state position versus membrane potential $n_\infty(E)$ can be described by Fermi distribution:

$$n_\infty = \frac{1}{1 - e^{-zF\left(E - E_{1/2}\right)RT}}.$$

The effective valency z of the gating charges was determined to about 4.5.

Several decades later, crystallographic structure analysis revealed that voltage-gated K$^+$ channels are formed by homotetramers, each subunit being composed of 6 transmembrane segments (TMS) (Fig. 6.8). Positively charge amino acid residues, in particular 4 arginines in the TMS 4, serve as the gating charges. Voltage-dependent conformational change involves movements of the TMS 6 and opening of the gate.

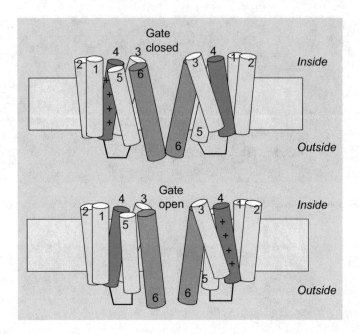

Fig. 6.8 Structure of a K$^+$ channel with 4 subunits (two are illustrated) with six transmembrane domains. The TMDs 4 with a series of positively charged residues represent the voltage sensor, the domains 6 the gate (based on Cha et al., 1999, Fig. 5, with kind permission from Springer Nature Publishing, 1999)

6.1.2.3 The Na$^+$ Channel

The kinetics of Na$^+$ channel gating with activation and inactivation is even more complex. There is an activation process that shows, qualitatively similar to the activation of K$^+$ conductance, a sigmoidal increase with time, which could be described by assuming three independent activation particles m. In addition, there is a slower inactivation process that shows simple exponential time course and, hence, was described by a single inactivation particle h. Only when all three m particles and the h particle are simultaneously in the open-state position, the channel is open, and the probability of being open is m^{3}h.

Hence the conductance was described by (compare Fig. 6.9):

$$g_{Na} = N \gamma_{Na} \ m^3 \ h = g^*_{Na} \ m^3 \ h$$

with maximum conductance g^*_{Na} and m and h having time and voltage dependence (Fig. 6.10) qualitatively similar to that described above.

At the resting potential the m-gates are predominantly closed (m $\approx$ 0), and the h-gate is with a probability of 70% open (h $\approx$ 0.7). In response to a depolarising pulse m increases very fast to a new potential-dependent steady state m$_\infty$, while h slowly decreases to a new potential-dependent steady state h$_\infty$.

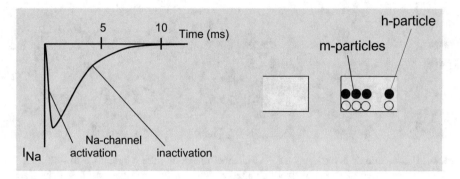

Fig. 6.9 HH model for the Na⁺ channel with 3 m and one h gating particles that can be in closed or open position

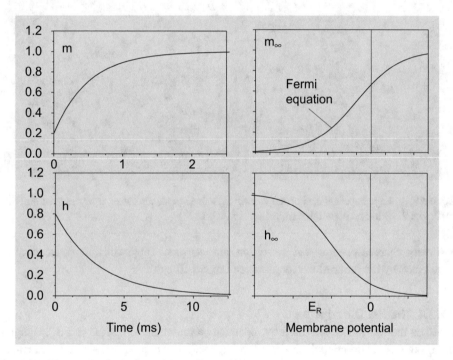

Fig. 6.10 Time and voltage dependence of m and h

The voltage dependencies of m_∞ and h_∞ can again be described by Fermi distribution with an effective valency of the gating charges for activation of about 6 and for inactivation of about 3.5.

The functional unit of the Na⁺ channel is a large α subunit with four repeats (Yu & Catteral, 2003), each composed of 6 TMS (Fig. 6.11); additional subunits have regulatory function. Like in voltage-gated K⁺ channels (Fig. 6.8) the TMSs 4 with a series of

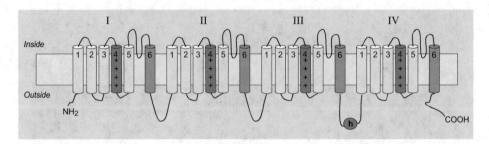

Fig. 6.11 Structure of a voltage-gated Na$^+$ channel illustrating the α subunit with 4 repeats. The TMSs 4 of each repeat with positively charged residues act as voltage sensor. h may serve as inactivation gate

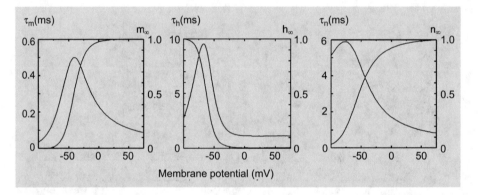

Fig. 6.12 Voltage dependencies of the τ_x and x_∞ values (x stands for m, h or n, based on Hille, 1970, Fig. 2, with kind permission from Elsevier AG, 1970)

positively charges residues act as the voltage sensors. The inactivation gate may be represented by the intracellular loop between repeat III and IV.

6.1.2.4 The HH Description

Based on the analysis of the carefully separated currents carried by Na$^+$ and K$^+$, Hodgkin and Huxley could describe the complete voltage and time dependencies of the currents by the m, h, and n values that were solutions of (see Sect. 6.1.2 (The Hypothetical Channel)):

$$\frac{dm}{dt} = \alpha_m(1-m) - \beta_m \qquad \frac{dn}{dt} = \alpha_n(1-n) - \beta_n n$$
$$\frac{dh}{dt} = \alpha_h(1-h) - \beta_h h.$$

An important advantage in the method was that they used rectangular voltage steps since the αs and βs then do not vary during the pulses. This allowed easy determination of the voltage dependence of the τ_x and x_∞ values (x stands for m, h or n) that are given by

$$\tau_x = \frac{1}{\alpha_x + \beta_x} \quad x_\infty = \frac{\alpha_x}{\alpha_x + \beta_x}.$$

Their voltage dependencies are shown in Fig. 6.12.

Though the data were obtained from completely separate current measurements, Hodgkin and Huxley were able to describe the total membrane current in response to a voltage step by summing up the different components:

$$j = C_m \frac{dE}{dt} + g_{Na}(E - E_{Na})m^3 h + g_K(E - E_K)n^4 + g_L(E - E_L).$$

This outcome illustrates the justification for the assumption that ion fluxes through the Na^+, K^+ and leak channels are independent from each other.

6.1.3 The Action Potential

6.1.3.1 Phenomenological Description

The time course of an action potential can be divided into four different phases (Fig. 6.13), and can be understood on the basis of the Hodgkin–Huxley description and the GHK equation for the potential.

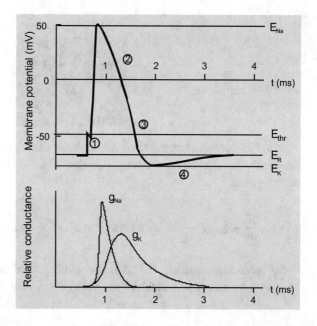

Fig. 6.13 Time course of potential and current during an action potential with depolarisation to the threshold potential E_{thr} and autoregenerative depolarisation to E_{Na} (1), Na channel inactivation (2), K channel activation, (3) and afterhyperpolarisation (4)

1. Depolarisation from the resting potential E_R (slightly positive to the Nernst potential E_K for K^+). If polarised beyond threshold E_{thr}, the potential responds with an overshoot approaching the Nernst potential E_{Na} for Na^+ due to depolarisation-induced Na^+ conductance.
2. Repolarisation due to spontaneous inactivation of Na^+ conductance.
3. Repolarisation speeded up by delayed activation of K^+ conductance.
4. Afterhyperpolarisation approaching the Nernst potential for K^+ due to elevated K^+ conductance compared to the resting state.

6.1.3.2 Calculation of Propagated Action Potential.

To describe the propagation of an action potential, Hodgkin and Huxley applied their model with the parameters determined under voltage clamp to the propagation of an undamped wave:

$$\frac{\partial^2 E}{\partial x^2} = \frac{1}{v^2}\frac{\partial^2 E}{\partial t^2}$$

The current flowing in a fibre segment Δx across the membrane is equal to the change in current flowing along the axon:

$$I = \frac{\pi a^2}{\rho}\frac{\partial E}{\partial x} \quad J = \frac{\delta I}{2\pi a \partial x} = \frac{a}{2\rho}\frac{\partial^2 E}{\partial x^2}$$

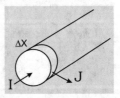

with I the current along the fibre, ρ in Ωcm, $2a$ fibre diameter, J density of current across the membrane. Combining the wave equation with the Hodgkin–Huxley equation for the current yields:

$$J = C_m\frac{dE}{dt} + g_{Na}(E - E_{Na})m^3 h + g_K(E - E_K)n^4 + g_L(E - E_L) = \frac{a}{2\rho}\frac{\partial^2 E}{\partial x^2} = \frac{a}{2\rho v^2}\frac{\partial^2 E}{\partial t^2}.$$

To solve this equation, Hodgkin and Huxley used a simple calculator to simulate iteratively an action potential. All the voltage and time dependencies of the parameters

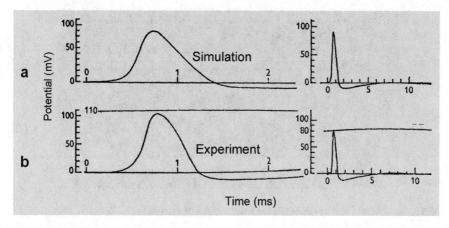

Fig. 6.14 Measured (**a**) and calculated (**b**) propagated action potentials shown for two different time scales (adapted from Hodgkin & Huxley, 1952, Fig. 15, with kind permission from John Wiley and Sons, 1952)

were those determined in voltage clamp, the only free parameter was the speed of propagation v. Estimates of v yielded that:

$$\text{for } v < v_{\text{measured}} : E \to +\infty$$
$$\text{for } v > v_{\text{measured}} : E \to -\infty$$

and only if v was chosen correctly, E followed the time course of an action potential. Figure 6.14. shows the result of their simulation (Fig. 6.14a) compared to their experimental data (Fig. 6.14b).

6.2   Continuous and Saltatory Spread of Action Potentials

6.2.1   The Electrotonic Potential

The description of the spread of an action potential by the wave equation is based on the all-or-none response, but also potential changes below threshold propagate along the fibre. To describe the spread of these so-called *electrotonic* potentials, we can treat the nerve fibre like a cable (Taylor, 1963) made up of an infinite sequence of four poles as illustrated in Fig. 6.15.

The section of an axon (Δx) is composed of the resistor of cytoplasm ($2R_{ax}$) and membrane (R_m), and of the membrane capacitance (C_m). The resistance of the external medium is small compared to R_{ax} and R_m, and hence will be neglected. For the limiting case of direct current flowing along the axon only ohmic resistances are of interest. Since the

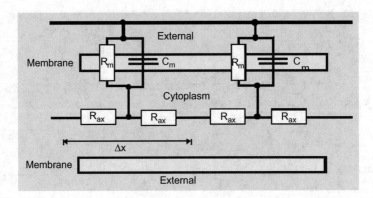

Fig. 6.15 Symbolised electric diagram for a section Δx of an axon

Fig. 6.16 Reduced electric
diagram of the axon for the case
of direct current. R_0 represents
the wave resistance of an infinite
cable

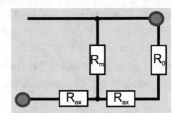

resistance of an infinite cable is not affected if a four pole is added, the resistance R_0 of the
cable can be calculated according to Fig. 6.16 with R_m parallel $R_{ax} + R_0$:

$$R_0 = R_{ax} + R_m \parallel (R_{ax} + R_0) \qquad \parallel \text{ denotes resistors in parallel} \qquad (6.1)$$

or

$$R_0 = R_{ax} + \frac{1}{\frac{1}{R_m} + \frac{1}{R_{ax} + R_0}}. \qquad (6.2)$$

This yields

$$R_0 = \sqrt{R_{ax}(R_{ax} + 2R_m)}. \qquad (6.3)$$

Since R_m is much larger than R_{ax}, we have approximately:

$$R_0 = \sqrt{2R_{ax}R_m}. \qquad (6.4)$$

Now we want to determine the change of potential along the axon. For this we consider
a cable element n preceding the resistance R_0 (Fig. 6.17) .

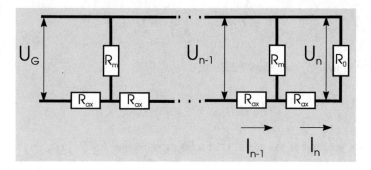

Fig. 6.17 Cable presentation of an axon composed of n segments

For the current in four poles $(n-1)$ we have

$$U_{n-1} = I_{n-1}R_0 \quad \text{and} \quad U_n = I_nR_0 \tag{6.5}$$

and for the potential drop U_{mn} across membrane resistance R_m of the nth segment:

$$U_{mn} = U_{n-1} - I_{n-1}R_{ax} = U_n + I_nR_{ax}. \tag{6.6}$$

With Eqs. (6.5) and (6.6) we obtain

$$U_{n-1} - U_{n-1}\frac{R_{ax}}{R_0} = U_n + U_n\frac{R_{ax}}{R_0}$$
$$U_n = U_{n-1}\left[\frac{1 - \dfrac{R_{ax}}{R_0}}{1 + \dfrac{R_{ax}}{R_0}}\right]. \tag{6.7}$$

The potential changes from four pole to four pole always by the same factor that is given by the expression in the []- brackets. Therefore, we can describe the potential at four pole n by

$$U_n = U_G\left[\frac{1 - \frac{R_{ax}}{R_0}}{1 + \frac{R_{ax}}{R_0}}\right]^n \tag{6.8}$$

with U_G representing the input potential. For the case of a nerve or muscle fibre the wave resistance R_o is large compared to R_{ax}, and hence we can write with $R_{ax}/R_o(<<1)$:

$$U_n \simeq U_G \left[1 - 2\frac{R_{ax}}{R_0}\right]^n \simeq U_G e^{-n\frac{2R_{ax}}{R_0}}.$$ (6.9)

Together with (6.4) this takes the form:

$$U_n \cong U_G e^{-\frac{n}{\sqrt{R_m/2R_{ax}}}} = U_G e^{-n/\lambda}.$$ (6.10)

This is the solution of the equation of a linear cable (see Eq. (6.11)):

$$\lambda^2 \frac{\partial^2 U}{\partial x^2} - U - \tau \frac{\partial U}{\partial t} = 0 \quad \text{with} \quad \partial U \partial t = 0$$

λ is termed length constant and is proportional to the square root of the diameter of the fibre ($\lambda \sim \sqrt{r}$):

$$\lambda = \sqrt{\frac{R_m}{R_i}} = \sqrt{\frac{\mathfrak{R}}{2\pi r} \frac{\pi r^2}{\rho_i}} \propto \sqrt{r}$$

with the membrane resistance $R_m = \mathfrak{R}/2\pi r$ given in Ωcm and the cytoplasmic resistance $R_i = 2R_{ax}$ in Ω/cm, λ has the dimension of a length. Typical values are in the range of 1–5 mm. For illustration, Fig. 6.18 shows the decline of potential along a nerve fibre next to an action potential assuming a length constant of 1 mm.

The eqs. (6.4) and (6.10) were derived for the case of direct current. Now we want to consider the case that a voltage U_G is set at $t = 0$ from 0 to U_o. Because of the capacitance the final value will not be reached instantaneously but will be approached asymptotically.

Fig. 6.18 Change of potential according to E = (50 mV-E_rest) · exp.(−x/1 mm) + E_rest along a cell fibre

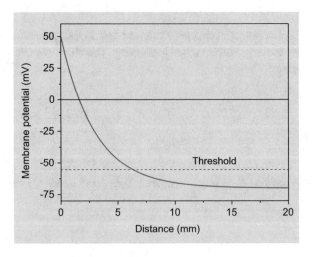

To describe this process the differential equation for a linear cable has to be solved (Hodgkin & Rushton, 1946; Taylor, 1963):

$$\frac{1}{R_{ax}} \frac{\partial^2 U}{\partial x^2} = C_m \frac{\partial V}{\partial t} + \frac{U}{R_m}.$$

(6.11)

Applying a rectangular voltage jump, the solution is

$$U(t) = U_o \frac{R_0}{R_v + R_0} erf \sqrt{\frac{t}{\tau}}$$

(6.12)

with $\tau = R_m C_m$ and the error function erf(x) is defined by

$$erf(x) = \frac{2}{\sqrt{\pi}} \int_0^x \exp\left(-y^2\right) dy$$

(6.13)

corresponds to the time, when U has reached about 84% of its final value; in other words at $erf(1.0) \cong 0.84$.

As an approximation, the potential change is usually described by an exponential time course $exp(t/\tau)$ with time constants τ in the range of 1–50 ms, depending on the membrane area and conductance. Figure 6.19 illustrates the change in membrane potential to a new steady state with time constant of 1 ms described by an exponential exp or an error function erf.

6.2.2 The Continuous Spread of an Action Potential

The spread of excitation depends on how fast the adjacent membrane areas can be depolarised beyond the threshold. This can be described on the basis of local currents that depolarise the neighbouring areas (see Fig. 6.20). The depolarisation in front of an

Fig. 6.19 Change of potential with time according to $\Delta E = E_1 + (E_0 - E_1) f(t/\tau)$

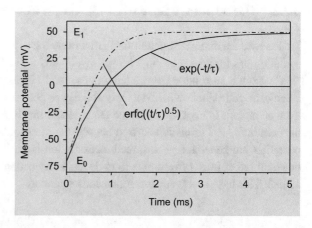

Fig. 6.20 Spread of excitation
by local currents

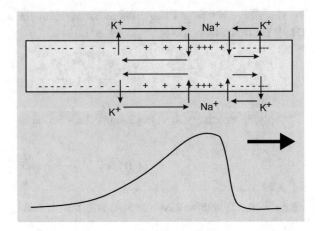

action potential mediated by K^+ outward current acts as the stimulus for generating action potential.

The velocity of spread of an action potential v is determined by λ and τ:

$$v = \lambda/\tau.$$

With otherwise constant electrical parameters the dependency of velocity on the radius of the fibre is given by

$$v \propto r^{0.5}.$$

To increase the velocity, evolution has created the giant nerve fibres of up to about 1 mm in diameter, which allows velocities up to 10 m/s. But this seems to be the upper limit. For mammalians, evolution has developed an additional variant of fast nerve fibres.

6.2.3 The Saltatory Spread of an Action Potential

To decrease the diameter but still further increase the velocity of spread of action potential, evolution has found a solution with the myelinated nerve fibres (Fig. 6.21).

The axon is surrounded over a length of several 100 µm by membrane layers formed by a Schwann cell (Virchow, 1854). Only at the gaps between the cells (nodes of Ranvier) the axon membrane is in contact with the extracellular medium (Ranvier, 1872). In these fibres the membrane has to be depolarised beyond the threshold only at the nodes where action potentials are generated. The spread of excitation is no longer continuous but the action potential rather jumps from node to node (saltatory action potentials (Huxley & Stämpfli, 1949)). This has several advantageous consequences:

Fig. 6.21 Schematic
myelinated nerve fibre

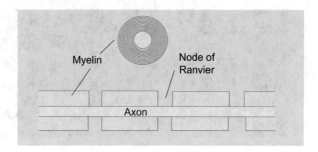

1. The speed of spread of an action potential is about 20x faster compared to a fibre of the same fibre diameter without myelination.
2. This allows that nerves can be thinner.
3. Since the generation of an action potential involves downhill movements of Na^+ and K^+ ions along their electrochemical gradients, the ions have continuously to be transported back by the Na^+, K^+ pump, which consumes metabolic energy in the form of ATP hydrolysis. Myelinated nerve fibres, where action potentials are generated only at the nodes, therefore, have a much lower consumption of metabolic energy.

6.3   Generation and Transmission of Action Potentials

6.3.1   Generation

Depending on the type of excitable cell, the mechanism of generating an action potential varies. One possibility are receptor cells receiving input from external signal, which they transduce into changes in membrane potential.

In a primary receptor cell, this polarisation can exceed threshold and elicits an action potential that spreads along its axon and is transmitted (see below) to another nerve cell. Secondary receptor cells transmit the polarisation to a nerve cell, where the received signal is summed up with other inputs.

Another possibility we have discussed already in the context of the heart muscle, where centres exist with spontaneously active cells. Figure 6.22 illustrates a possible mechanism, which involves a voltage-dependent Ca^{2+} conductance (g_{Ca}) and a Ca^{2+}-activated K^+ conductance (g_K).

In small cells opening of Ca^{2+}-selective channels by depolarisation may result in an increase of Ca^{2+}_i that in turn activates the K^+ conductance facilitating repolarisation. This leads to closing of the Ca^{2+} channels and Ca^{2+}_i-regulating mechanisms re-establish sub-micromolar activity. The resulting closing of the K^+ channels will again depolarise the membrane, and the cycle will start again. The above described mechanism has been discussed for certain neurons; in heart different mechanisms are involved, but also the interplay with Ca^{2+} channels plays a key role.

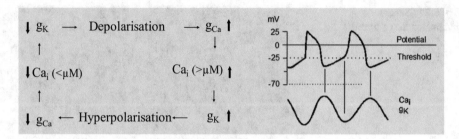

Fig. 6.22 Example for the mechanism of spontaneous electric activity

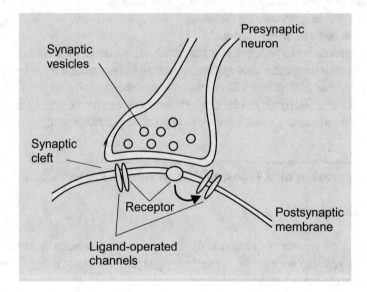

Fig. 6.23 Schematic drawing of a synapse

6.3.2   Transmission

The dominating mechanism for transmission of electrical signals from one cell to another one is the synaptic transmission (Fig. 6.23). The ionic basis of this signal transmission was first discovered by Eccles (Eccles, 1963), for which he was honoured together with Hodgkin (Hodgkin, 1963) and Huxley (Huxley, 1963) with the Nobel prize in 1963. The transmission between nerve and peripheral cells, like muscle cells, was further elaborated by Katz (Katz, 1970), (v. Euler, 1970), and (Axelrod, 1970) and rewarded with Nobel prize in 1970.

On the arrival of an action potential at the ending of the presynaptic nerve cell Ca^{2+} entry triggers the release of a transmitter substance from the synaptic vesicles to the synaptic cleft (see also Fig. 7.12). The neurotransmitter diffuses to the opposing membrane of the postsynaptic cell and modulates through specific receptors the activity of ion-selective channels. As a consequence, the membrane potential changes. Depending on the type of channel, the synapse is excitatory and the membrane potential depolarises or the synapse is inhibitory and the membrane potential hyperpolarises. While GABA (γ-amino buteric acid) is the dominating inhibitory transmitter, glutamate is the dominating excitatory transmitter. To terminate neurotransmission the respective neurotransmitter is removed from the synaptic cleft by neurotransmitter uptake transporters (see Sects. 7.1.5, 8.4.2, 8.5.2).

6.4 Summary of the Different Types of Potentials

In the following table the different versions of membrane potential we have discussed so far are summarised (Table 6.2).

6.4.1 Surface Potential

The listing of membrane potentials should be completed by a short description of surface potentials existing at all cell membranes.

At the surface of a cell membrane negative charges are dominating. This leads to surface potentials that alter the electrical field in the membrane (Fig. 6.24, for details see McLaughlin, 1989).

For a given surface charge density σ, the surface potential ΔE_o depends on electrolyte concentration c and ion charge z, and can be described by the Graham equation:

$$\sigma^2 = 2\varepsilon\varepsilon_0 RT \sum_k c_k \left(e^{-z_k F\Delta E_0/RT} - 1 \right)$$

which is based on Gouy–Chapman theory assuming a planar, continuous surface charge density. The surface charges are screened by the ions in the electrolyte solution leading to

Table 6.2 Listing of various membrane potentials (the last column refers to the respective section)

Nernst potential	$E_{rev} = -\frac{RT}{F} \ln \left(\frac{[X]_i}{[X]_o} \right)$	2.3.2
Donnan potential	$E_d = -(RT/F) \ln \left([K_I]/[K_O] \right)$ $= -(RT/F) \ln \left[\frac{[A]}{2[K_o]} + \left(\left(\frac{[A]}{2[K_o]} \right)^2 + 1 \right)^{1/2} \right]$	2.3.1
Reversal potential	$E_{GHK} = E_{rev} = \frac{RT}{F} \ln \left(\frac{P_{Na}[Na]_o + P_K[K]_o + P_{Cl}[Cl]_i}{P_{Na}[Na]_i + P_K[K]_i + P_{Cl}[Cl]_o} \right)$	2.4
Electrotonic potential	$E(x) = E_0 e^{-x/\lambda}$ $E(t) = E_\infty + (E_\infty - E_0) erfc \left(\sqrt{t/\tau} \right)$	6.2.1
Action potential	$E(t)$ as solution of $J = \frac{a}{2\rho v^2} \frac{\partial^2 E}{\partial t^2}$	6.1.3
Threshold potential	$E_S = E(I_{Na} = I_K)$	6.1.3

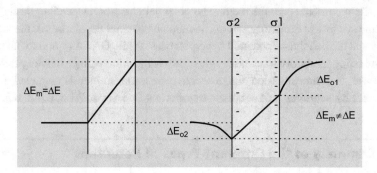

Fig. 6.24 Effect of surface charge on the electrical field in the membrane

the formation of an electrical double layer. ΔE_o has the same sign as σ and decreases with increasing ion concentration.

For small surface potentials ($\Delta E_o < <RT/F$) we have

$$\Delta E_0 = \frac{\sigma r_D}{\varepsilon \varepsilon_0} \quad \text{with Debye length} \quad r_D = \frac{1}{zF}\sqrt{\frac{\varepsilon \varepsilon_0 RT}{2c}}$$

$$\Delta E_0(x) = \Delta E_0 e^{-xr_D}.$$

If several ion species contribute, the concentration c has to be replaced by the ionic strength I:

$$I = \frac{1}{2}\sum_k z_k^2 c_k \quad \text{and hence} \quad r_D = \frac{1}{F}\sqrt{\frac{\varepsilon \varepsilon_0 RT}{2I}}.$$

Due to the surface potential, the concentration of an ion species c_{k0} differs from the bulk concentration c_k:

$$c_{k0} = c_k e^{-z_k \Delta E_0 F RT}.$$

At a surface potential of -40 mV monovalent cation concentration would be 4.5 times higher at the surface than in the bulk solution, and for divalent cation the concentration would be even 20 times higher. Even at -10 mV divalent cation concentration would be more than two times higher.

6.5 Action Potential in Non-nerve Cells

In the following we will briefly illustrate the various action potentials that can be detected in different types of cells.

6.5.1   Skeletal Muscle

The action potential that can be recorded from a skeletal muscle fibre is composed of two components (see Fig. 6.25) that originate from the action potential spreading along the muscle fibre surface and from the action potential spreading into the T-tubular system, which governs the release of Ca^{2+} from the sarcoplasmic reticulum and hence contraction.

Like in nerve cells, the ionic basis of the action potential in a skeletal muscle fibre is the activation and spontaneous inactivation of Na^+ channels and activation of K^+ channels.

6.5.2   Smooth Muscle

In smooth muscle cells, fast-gated Ca^{2+} channels play the same role as the Na^+ channels play in nerve and skeletal muscle fibres though on a slower time scale (Tomita & Iino, 1994). In addition, slow-gated Ca^{2+}- and Na^+-permeable channels shape the time course of the action potential (Fig. 6.26).

The overlapping of the channel gating with intracellularly elevated Ca^{2+} activity allows the constant tonus of muscle fibres.

6.5.3   Heart Muscle

The heart muscle is composed of different types of cells. In the working myocard, the action potential is mainly governed by Na^+ and Ca^{2+} channels that have qualitatively similar kinetics with activation and inactivation although the Ca^{2+} channels are much slower (Fig. 6.27).

Fig. 6.25 Action potential in skeletal muscle and associated changes in conductance

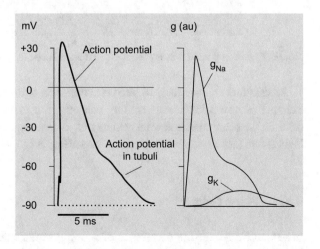

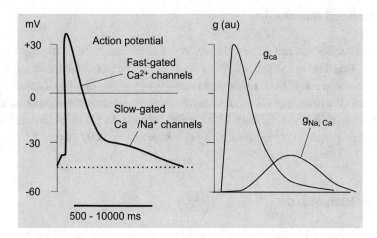

Fig. 6.26 Action potential in smooth muscle and associated changes in conductance

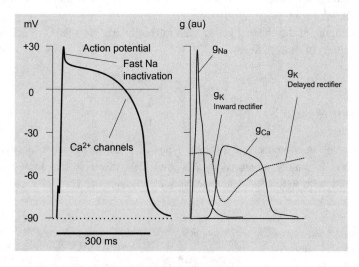

Fig. 6.27 Action potential in myocardium and associated changes in conductance

An essential characteristic of heart function is its autonomous spontaneous activity. An example of how spontaneous activity can originate in neurones was described in Sect. 6.3.1. In the heart muscle several centres of spontaneous activity exist. The centre with the highest rate (see Fig. 6.28) is the Sinus node (Fig. 3.2).

6.5.4   Plant Cells

In several plant cells the occurrence of action potentials has been demonstrated (Wayne, 1994). Examples are mimosa, Venus fly-trap and algae. In many cases protoplasma

Fig. 6.28 Pacemaker action
potential in the Sinus node

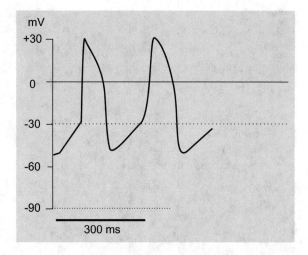

Fig. 6.29 Action potential in a
plant cell and associated changes
in conductance

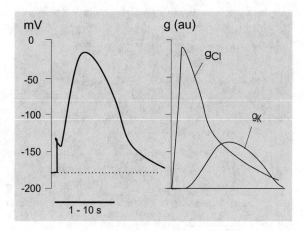

streaming is blocked on the occurrence of an action potential. A schematic illustration of an
action potential in a plant cell is given in Fig. 6.29.

Qualitatively, the time course is similar to that of an action potential in a nerve or muscle
cell, except that the processes are much slower. Instead of the activation and inactivation of
Na^+-inward flux Cl^- efflux occurs. Interestingly to mention, the resting potential of a plant
cell can be extremely negative and reaches values close to -200 mV. This negative
potential is generated by the activity of an electrogenic ATPase transporting H^+ across
the membrane.

Take-Home Messages

1. **The HH description of an action potential** is based on **time- and voltage-dependent ion-specific permeabilities**/conductances

$$I_m(t,E) = C\frac{dE}{dt} + I_{\text{Leak}}(E) + I_{\text{Na}}(t,E) + I_K(t,E).$$

2. **The voltage-dependent gating** of the ion channels is described by potential-dependent rate constants between open and closed states of gating particle x (m, h, n)

$$\frac{dx}{dt} = \alpha(1-x) - \beta x \quad x(t) = x_\infty - (x_\infty - x_0)e^{-t/\tau}$$

$$\text{with} \quad x_\infty = \frac{\alpha}{\alpha + \beta} \qquad \tau = \frac{1}{\alpha + \beta}$$

   leading to the time- and voltage-dependent conductances:

$$g_{\text{Na}} = N\gamma_{\text{Na}}\,m^3\,h \quad \text{and} \quad g_K = N\,\gamma_K\,n^4.$$

3. **The time course of an action potential** is governed by
   (a) fast **regenerative depolarisation** when a threshold potential is exceeded due to **fast activation** (m particles) **of g_{Na}**.
   (b) slower **repolarisation** due to **slower inactivation** (h particles) **of g_{Na}**.
   (c) **speeded-up repolarisation** due to **slow activation** (n particles) **of g_K** resulting in an **after hyperpolarisation**.
   (d) **return to resting potential** due to g_K returning to resting value.

4. **The propagation of an action potential** is described by an undamped wave:

$$J = C_m\frac{dE}{dt} + g_{\text{Na}}(E - E_{\text{Na}})m^3h + g_K(E - E_K)n^4 + g_L(E - E_L) = \frac{a}{2\rho}\frac{\partial^2 E}{\partial x^2} = \frac{a}{2\rho v^2}\frac{\partial^2 E}{\partial t^2}.$$

5. Action potentials can spread along a cell fibre as a **continuous wave with speed v**:

$$v = \lambda/\tau \propto r^{0.5}.$$

   The speed can be improved by **saltatoric** (from node of Ranvier to node of Ranvier) conduction allowing
   (a) faster signal conduction.
   (b) thinner cell fibres.
   (c) less metabolic energy consumption.

(continued)

6. The dominating mechanism for **transmission of electrical signals** from one cell to another one is **synaptic transmission** with transmitter release at the presynaptic cell and activation of excitatory or inhibitory receptors channels at the postsynaptic cell.
7. The **triggering of an action potential** is achieved by modulation of membrane potential by
    (a) activation of specific receptors.
    (b) hormones and transmitters.
    (c) spontaneous activity.
8. Also without voltage- and time-dependent conductances an applied change of membrane potential spreads in time (error function) and space (exponential), in form of the so called **electrotonic potential**.
9. The cell membranes with their phospholipids carry negative surface charges, which result in a **surface potential** that significantly influences the potential within the membrane and local ion activities.

Exercises

1. What was the theoretical, methodological and experimental basis of the work of Hodgkin and Huxley?
2. Describe the voltage and time dependence of membrane current on the basis of the Hodgkin–Huxley description.
3. Describe the propagation of an action potential on the basis of Hodgkin–Huxley description.
4. Describe the space and time dependency of potential propagation along a linear cable.
5. What are the advantages of myelinated nerve fibres?
6. Describe a mechanism of spontaneous activity.
7. Describe the mechanism of synaptic transmission.
8. What are the consequences of Na^+-channel inactivation?
9. How can a surface potential be determined, and described consequences?
10. Describe difference of action potentials in different cell types and their basis in terms of ion permeabilities.

References

Axelrod, J. (1970). Noradrenaline: fate and control of its biosynthesis. *Nobel Lectures 1963–1970*
Bernstein, J. (1902). Untersuchungen zur Thermodynamik der bioelektrischen Ströme. *Archiv für die gesamte Physiologie des Menschen und der Tiere, 92*, 521–562.

Bernstein, J. (1912). *Elektrobiologie*. Vieweg.

Cha, A., Snyder, G. E., Selvin, P. R., & Bezanilla, F. (1999). Atomic scale movement of the voltage-sensing region in a potassium channel measured via spectroscopy. *Nature, 402*, 809–813.

Eccles, J. C. (1963). Noradrenaline: fate and control of its biosynthesis. *Nobel Lectures 1963–1970*

Euler, U. (1970). Adrenergic neurotransmitter functions. *Nobel Lectures 1963–1970*

Hille, B. (1970). Ionic channels in nerve membranes. *Progress in biophysics and molecular biology, 21*, 1–32.

Hille B. (1992, 2001) *Ionic channels of excitable membranes*, Sinauer Associates Inc.,

Hodgkin, A. L. (1963). The ionic basis of nervous conduction. *Nobel Lectures 1963–1970*

Hodgkin, A. L., & Huxley, A. F. (1952). A quantitative description of membrane current and its application to conductance and excitiation in nerve. *The Journal of Physiology, 117*, 500–544.

Hodgkin, A. L., & Katz, B. (1949). The effect of sodium ions on the electrical activity of the giant axon of the squid. *The Journal of Physiology (Lond), 108*, 37–77.

Hodgkin, A. L., & Keynes, R. D. (1955). The potassium permeability of a giant nerve fibre. *The Journal of Physiology, 128*, 61–88.

Huxley, A. F. (1963). The quantitative analysis of excitation and conduction in nerve. *Nobel Lectures 1963–1970*

Huxley, A. F., & Stämpfli, R. (1949). Evidence for saltatory conduction in peripheral myelinated nerve fibres. *The Journal of Physiology, 108*, 315–339.

Katz, B. (1970). On the quantal mechanism of neural transmitter release. In *Nobel Lectures 1963–1970*.

McLaughlin, S. (1989). The electrostatic properties of membranes. *Annual Review of Biophysics and Biophysical Chemistry, 18*, 113–136.

Neher, E., & Sakmann, B. (1976). Single-channel currents recorded from membrane of denervated frog muscle fibres. *Nature, 260*, 799–802.

Ranvier, L.-A. (1872). Recherches sur l'histologie et la physiolige des nerfs. *Archives Physiology Normal Pathologique, IV/2*, 129–149.

Taylor, R. E. (1963). Cable Theory. In W. L. Nastuk (Ed.), *Physical Techniques in Biological Research* (pp. 219–262). Academic Press.

Tomita, T., & Iino, S. (1994). Ionic Channels in Smooth Muscle. In L. Szekeres & J. G. Papp (Eds.), *Pharmacology of Smooth Muscle* (Handbook of Experimental Pharmacology) (Vol. 111, pp. 35–56). Springer.

Virchow, R. (1854). Über das ausgebreitete Vorkommen einer dem Nervenmark analogen Substanz in dem tierischen Gewebe. *Archiv für pathologische Anatomie und Physiologie und für klinische Medicin, 6*, 562–572.

Wayne, R. (1994). The excitability of plant cells: with a special emphasis on characean internodal cells. *The Botanical Review, 60*, 265–367.

Yu, F. H., & Catteral, W. A. (2003). Overview of voltage-gated sodium channel family. *Genome Biology, 4*, 207.

Carrier-Mediated Transport

7

Contents

Abstract

The sensitivity of modern voltage-clamp techniques allows detection of carrier-mediated transport provided that a large membrane surface is available. *Xenopus* oocytes are as an expression system an ideal preparation. Chapter 7 shall illustrate by three examples, the anion exchanger, the Na,K pump and the GABA transporter, characteristics of carriers compared to channels and how electrophysiological methods can be used for functional characterisation.

Keywords

Carrier proteins · Anion exchanger · Na,K pump · Neurotransmitter transporter · Access channel

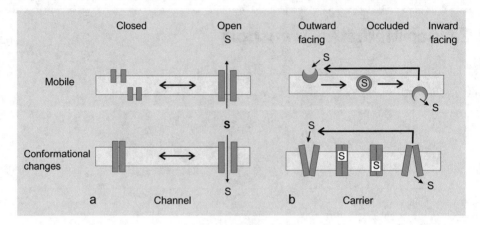

Fig. 7.1 (**a, b**) Cartoon on how pores (**a**) and carriers (**b**) can mediate transport either by structures floating in the membrane (upper panel) or by transmembrane proteins (lower panel)

7.1 General Characteristics of Carriers

7.1.1 Distinction Between Pores and Carriers

The channels or pore-forming proteins, in their simplest version, exist in two principally different forms; one, in which they do not form a pore (the closed form), and in an open form, in which a porous structure spans the entire membrane. Such an open pore allows more or less free diffusion of ions along their electrochemical gradient. Pores can be created in two different ways as illustrated in Fig. 7.1a, either the channel can be formed by floating subunits (e.g. half pores) that can associate to form a transmembrane pore or a pre-existing transmembrane protein undergoes conformational changes between a closed and an open form (for characteristics of pores compared to carriers, see Sect. 1.1, Fig. 1.2).

The function of pores can be described by reaction diagrams for the gating, where the different states represent different open and closed conformations of the channel protein. Consequently, the transitions represent the opening and closing of the pores. As an example, we discussed already the HH-model (Sect. 6.1.2) with the closed resting (R), the open activated (O) and the closed inactivated states (I). A possible reaction diagram would be.

Similar to what we discussed for pore formation, carrier transport can be mediated by a mobile transport molecule or a pre-existing transmembrane protein that undergoes conformational changes with inward and outward orientated binding sites (see Fig. 7.1b). For carriers the mechanism of transport can also be described by reaction diagrams for transitions between different conformations of the carrier molecule. Here the different conformations represent different states of substrate interaction with the transport protein, and the transitions represent the actual transport of the substrate across the membrane. In the simplest way, transport involves binding of the substrate on one side, translocation across the membrane, and release on the other side. A simple example would be:

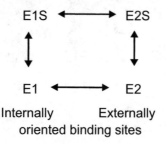

Outward transport of substrate will be described by cycling clock-wise through the above diagram.

For natural transport proteins in a cell membrane, only the transmembrane versions exist. But there exists also a variety of antibiotics where mobile subunits interact with each other to form a pore like gramicidin A or alamethicin or where a mobile molecule diffuses in the membrane to mediate transport like valinomycin or nonactin.

If net charges are transported across the membrane, a current is generated. Consequently, the membrane potential contributes to the driving force. Thermodynamically, a potential E_r should exist where the current vanishes. In case that substrates are on both sides of the membrane, a reversal potential can be expected. For channels, E_r is given by the Nernst potential for the permeating ion; in case of several permeating ion species with the same absolute valency z, the GHK equation may be used (Sect. 2.4):

$$E_r = E_{\text{GHK}} = -\frac{RT}{zF} \ln \frac{\sum P_k [c_k]_o}{\sum P_k [c_k]_i}.$$

For a carrier transporting n substrate molecules and z net charges across the membrane, coupled to m_k ions of species c_k, we have under steady-state conditions for the electrochemical potential μ_S:

$$\mu_S = \sum \mu_k + zEF \quad \text{or}$$
$$\ln \left(\frac{[S]_i}{[S]_o} \right)^n = \sum_k \ln \left(\frac{[c_k]_o}{[c_k]_i} \right)^{m_k} + \frac{zFE_r}{RT}.$$

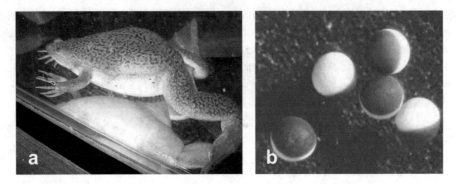

Fig. 7.2 The clawed toad *Xenopus laevis* (**a**), full-grown, prophase-arrested oocytes (**b**)

Hence the reversal potential for such a carrier is.

$$E_r = -\frac{RT}{zF}\left(\ln\frac{[S]_i}{[S]_o} + \sum \ln\left(\frac{[c_k]_o}{[c_k]_i}\right)^{m_k}\right).$$

In the following we will present a general view on how carrier transport works, and how we can use electrophysiology to investigate the mechanisms of carrier transport and their role in physiological and pathophysiological processes.

7.1.2   The Oocytes of *Xenopus*: A Model System

A major problem in the electrophysiological investigation of carrier transport is the low rate of transport (comp. Fig. 1.2). Therefore, to investigate carrier transport, a cell with a large surface area and a high density of transporters is essential. In addition, low background conductances from open channels are helpful. The oocytes of the African clawed toad *Xenopus laevis* (Fig. 7.2) have turned out to be an ideal model system for electrophysiological characterisation of particularly carrier transport.

These cells have a diameter of more than 1 mm and a surface area of several mm². They can easily be impaled with microelectrodes; the two-electrode voltage clamp (Sect. 4.1.3), the cut-oocytes voltage clamp (Sect. 4.1.5), and all excised versions of patch clamp (Sect. 4.4) can be applied. It is even possible to impale the membrane with additional microelectrodes to monitor, e.g. intracellular ion activities with ion-selective microelectrodes (Sect. 3.4.3).

Due to the large volume of the oocyte, concentrations in the cytoplasm of an intact cell do hardly change during an experiment, but temporary pre-treatment of the oocyte with Ca^{2+}-free solution makes the membrane permeable for Na^+ and K^+. This can be used to change the intracellular Na^+ and K^+ concentration of the oocyte before an electrophysiological experiment. Figure 7.3 illustrates the changes in ion activities as an example for the application of ion-selective microelectrodes.

Fig. 7.3 Intracellular Na$^+$ and K$^+$ activities in *Xenopus* oocytes measured with ion-selective microelectrodes during treatment with Ca^{2+}-free solution (Silke Elsner and Wolfgang Schwarz, unpublished)

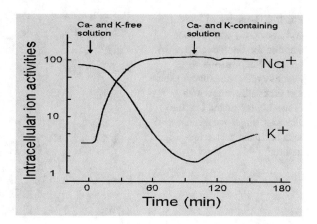

In addition to their ideal accessibility to electrophysiological techniques, the oocytes can be used as an expression system. Again, it is their large size which allows microinjection of DNA or RNA into the cell, and they have the complete machinery to process the DNA or mRNA to synthesise the coded proteins and to incorporate them into their own cell membrane, where their function can be studied. The large size of the cells even allows to apply biochemical methods to a single cell. We will show in the following by three examples how electrophysiology can be used in this sense, and in particular how electro-physiology can be combined with biochemical, molecular biological, and pharmacological techniques to address basic questions of transport function and structure–function relations and of medical, pharmacological interest.

Due to their easy handling and accessibility to electrophysiological methods, the *Xenopus* oocytes have become a model system also for the investigation of expressed channel proteins. We will first present three examples of carriers to illustrate typical characteristics that can be investigated by electrophysiological techniques.

7.1.3   The Anion Exchanger

The anion exchanger of red-blood cells, also called band-3 protein, mediates exchange of Cl$^-$ by HCO$_3^-$ across the cell membrane. The transporter has been studied for the longest time period and has been characterised by biochemical methods and flux measurements (Passow, 1986). Its physiological function and the reaction diagram are illustrated in Fig. 7.4.

The physiological role is to increase the transport capacity of the blood for CO$_2$. This is achieved by the carboanhydrase inside the red-blood cell, which readily transfers the CO$_2$ to HCO$_3^-$ and vice versa. Under physiological conditions the transporter operates in a HCO$_3^-$/Cl$^-$ exchange mode; under experimental conditions it can operate also as Cl$^-$/Cl$^-$ exchanger. In the E1 conformation, an intracellular anion A can bind followed by confor-mational change to E2 with extracellular release. Only after rebinding of an extracellular anion, the conformational transition back to E1 can occur. This mode of transport is

Fig. 7.4 Function of band-3 in
erythrocyte (**a**), and reaction
diagram for anion exchange (**b**)
with E1 conformation with
inwardly orientated binding sites
and E2 conformation with
outwardly orientated binding
sites (for simplicity the states
with occluded anions are
omitted)

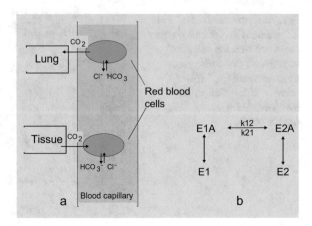

electrically silent, but nevertheless, could be modulated by membrane potential, e.g. if
charges are moved within the protein during conformational changes, which can be
described by voltage-dependent rates k_{12} and k_{21}:

$$k_{12} = k_{12}^0 e^{-vEF/RT} \quad \text{and} \quad k_{21} = k_{21}^0 e^{+uEF/RT}.$$

To study whether the anion exchange is voltage-dependent, we have to perform
measurements of tracer fluxes at different membrane potentials. Of course, this is not
possible with the tiny single red-blood cells. But it is possible to use the *Xenopus* oocytes
with expressed anion exchanger. This type of experiment is illustrated in Fig. 7.5a.
Radioactivity in ^{36}Cl$^-$-loaded oocytes is recorded on a Geiger–Müller tube under voltage
clamp; the exponential decline of radioactivity during perfusion of the chamber
(Fig. 7.5a) reflects release of Cl$^-$. The contribution by the anion exchanger is determined
by the difference of rate of release in the absence and presence of a specific inhibitor. The
band-3-mediated Cl$^-$ release is voltage-dependent (Fig. 7.5c) despite of its electrically
silent transport mechanism. If the voltage dependence of Cl$^-$ efflux would be due to
voltage-dependent conformational changes (voltage-dependent k12 and k21, Fig. 7.4b) a
maximum would be expected. The absence of a maximum might be attributed to an
extreme asymmetric distribution in favour of, e.g. E1 (see Schwarz et al., 1992 and
Fig. 7.6).

For the simulation of the data, a slight additional voltage dependence can be assigned to
external anion binding. Such voltage-dependent ion interaction can be interpreted by an
access channel within the electrical field that has to be passed before the ion can reach its
binding site (see Fig. 7.6).

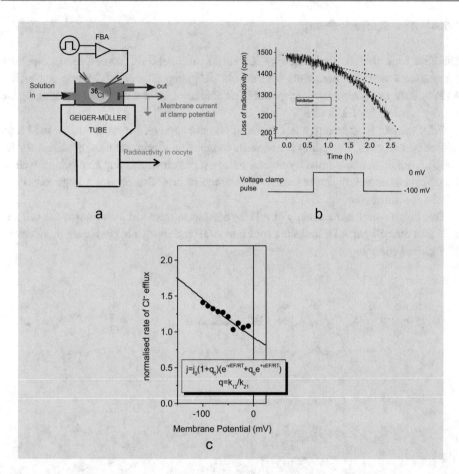

Fig. 7.5 Illustration (**a**) of flux measurement under voltage clamp in oocytes with expressed band-3, and (**b**) of a typical recording of exponential loss of radioactivity, and (**c**) of the results of voltage-dependent rate of band-3-mediated Cl^- release (according to Grygorczyk et al., 1987, Fig. 1 and 8, with kind permission from Springer Nature, 1987)

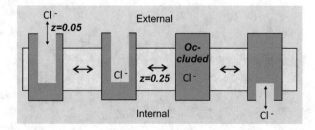

Fig. 7.6 Cartoon of voltage-dependent binding and conformational changes (E1 $\leftrightarrow$ E2) of the band-3 anion exchanger. z values represent effective valencies of charges moved during steps associated with external Cl^- binding and conformational changes, respectively

7.1.4 The Sodium Pump

The first time that the existence of an access channel in a carrier transporter was demonstrated was in experiments with the Na^+,K^+ pump. The Na^+,K^+ pump or Na^+,K^+-ATPase is the most important transporter in all animal cells responsible for maintaining the activity gradient for Na^+ and K^+.

Cells without functioning pump, therefore, cannot survive. An exception are red-blood cells of certain dogs. The pump is necessary to maintain electrochemical gradients for K^+ and Na^+, which are consumed by several transport proteins including channels as well as the so-called secondary active carriers and which in turn also control a large variety of cellular functions (see Fig. 7.7).

This pump uses the free energy of ATP hydrolysis to transport 3 Na^+ out of the cell and 2 K^+ into the cell per ATP molecule split into ADP and inorganic phosphate by releasing 30.5 kJ/mol energy:

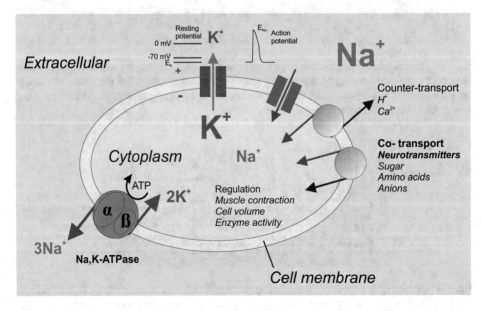

Fig. 7.7 Functional significance of the Na^+,K^+ pump for maintaining Na^+ and K^+ gradients

7.1.4.1 Steady-State Pump Current

As a consequence of the $3Na^+, 2K^+$ stoichiometry, the transporter is electrogenic, i.e. it generates a current. This current can serve as a measure of transport activity and (at least in principle) can be detected under voltage clamp. The reaction diagram of the Na^+, K^+ pump is much more complex (see Fig. 7.8) than the diagram we used for description of anion exchanger (compare Fig. 7.4b).

Due to its electrogenicity, the membrane potential adds to the driving force and makes pump activity for thermodynamic reasons potential-dependent. The pump currents under physiological conditions show pronounced voltage dependency with a maximum occurring under physiological conditions (compare Fig. 7.9 right), which suggests two voltage-dependent steps. The voltage dependence strongly depends on the extracellular Na^+ and K^+ concentrations. A straight-forward explanation, similar to the explanation for the anion exchanger, would be the existence of an external access channel for both Na^+ and K^+ binding because the voltage dependencies strongly depend on external Na^+ and K^+ with effective valencies z_K and z_{Na} (see Fig. 7.10), which are a measure for the apparent dielectric length of the access channels. These values can be determined from the potential dependence of the pump current at different K^+ or Na^+ concentrations. For a complete description an additional potential-dependent contribution has to be added, and may possibly reflect potential-dependent conformational changes, similar to what we have

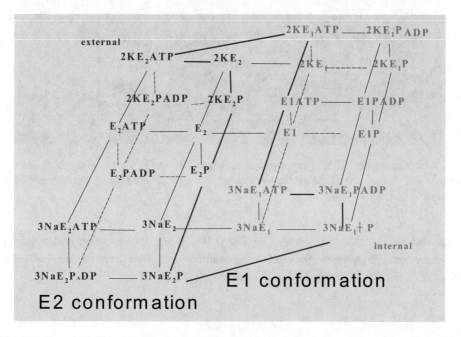

Fig. 7.8 Reaction diagram of Na^+, K^+ ATPase with E_1 conformations (internal-oriented cation binding sites) and E_2 conformations (external-oriented binding sites) (based on (Vasilets & Schwarz, 1993), Fig. 2, with kind permission from Elsevier AG, 1993)

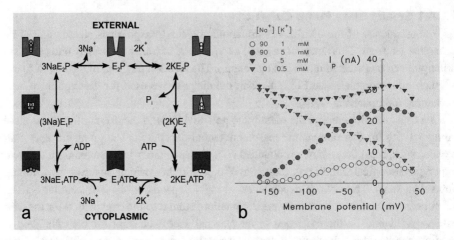

Fig. 7.9 (a) Reduced reaction diagram of the Na⁺,K⁺-ATPase, and (b) voltage dependencies of pump current under different external Na⁺ and K⁺ concentrations (based on Vasilets and Schwarz (1993), with kind permission from Elsevier AG, 1993)

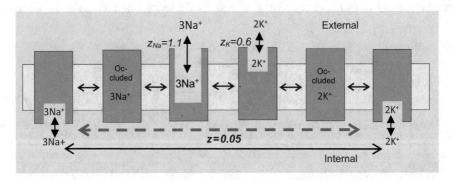

Fig. 7.10 Cartoon for Na⁺,K⁺ Pump with access channel Na⁺ and K⁺). z values represent effective valencies of charges moved during steps associated with external ion binding and conformational changes (E1 ↔ E2), respectively

discussed for the anion exchanger (see Fig. 7.10). While for the anion exchanger the conformational change gives the dominating contribution, for the pump the external cation binding dominates the potential dependence.

7.1.4.2 Transient Pump-Generated Currents

If we interrupt the transport cycle by removal of external K⁺, the transporter can perform Na⁺,Na⁺ exchange (left branch in reaction diagram, Fig. 7.9a), and we can analyse partial reactions:

Fig. 7.11 Transient currents in Na⁺/Na⁺ exchange mode described by the sum of three exponentials (according to Salonikidis et al., 2000, with kind permission from Elsevier AG 2000)

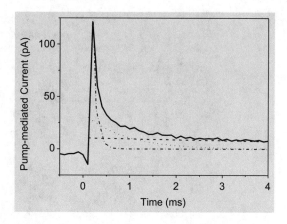

$$\text{E1ATP} \rightleftharpoons \text{3NaE1ATP} \rightleftharpoons \text{(3Na)E1P} \rightleftharpoons \text{3NaE2P} \rightleftharpoons \text{E2P}$$

If a potential step is applied to a membrane, the voltage dependence of the transition rates will lead to a new distribution between the different states in the reaction diagram, and the associated charge movements will give rise to transient currents (see Fig. 7.11). Because of the large surface area of an oocyte, charging the membrane capacitance takes too much time to accurately record such transient currents, but this is possible with the giant-patch technique (see Sect. 4.4). Such transients provide kinetic information about single steps in the reaction cycle. The different components possibly reflect two voltage-dependent binding steps and a voltage-dependent conformational change.

7.1.5   The Neurotransmitter Transporter GAT1

On the arrival of an action potential at the ending of a presynaptic nerve fibre Ca^{2+} enters the cell initiating vesicular release of a neurotransmitter; the transmitter activates ionotropic and G-protein-coupled receptors at the postsynaptic membrane (see Fig. 7.12). The activation is terminated by removal of the neurotransmitter by secondary active Na^+-gradient-driven neurotransmitter uptake transporters. As an example, for a secondary active transporter, we will briefly deal with the GABA transporter. GABA is the dominating transmitter an inhibitory synapse in the central nervous system.

The transporter named GAT mediates the uptake of 1 GABA by simultaneous inward movement of 2 Na^+ and 1 Cl^-. The energy liberated by the movement of Na^+ and Cl^- along their electrochemical gradients is used for the uphill transport of GABA. The corresponding reaction diagram is illustrated in Fig. 7.13a.

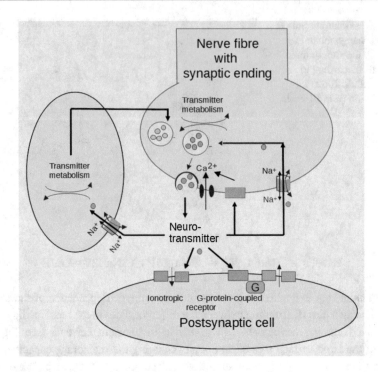

Fig. 7.12 Synaptic transmission: Vesicular release of neurotransmitter by the presynaptic neuron, its binding to postsynaptic receptors and its removal by Na^+-driven neurotransmitter transporters

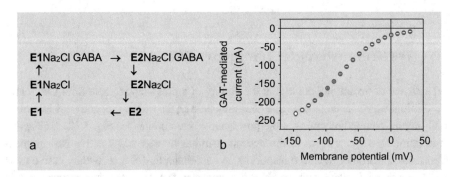

Fig. 7.13 Reaction diagram of GABA transporter GAT1 (**a**) and typical voltage dependence of GAT1-mediated current determined as GABA-induced current in *Xenopus* oocytes (**b**)

Due to the $2Na^+/1Cl^-$ stoichiometry, the transporter is electrogenic generating an inward-directed current. As consequence, the membrane potential adds to driving force, and the GAT-mediated current will exhibit voltage dependence as illustrated in Fig. 7.13b.

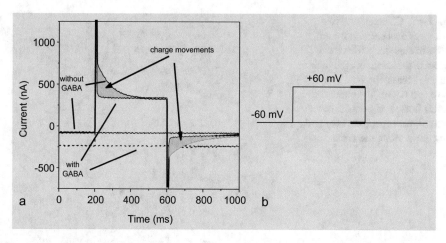

Fig. 7.14 Current signals (**a**) in response to rectangular voltage steps (**b**) in the absence and presence of GABA. (based on Eckstein-Ludwig et al., 1999, Fig. 1, with kind permission from John Wiley and Sons, 1999)

The GAT-mediated current can be obtained from the difference of current signals in the presence and absence of extracellular GABA (see Fig. 7.14). As expected, at the negative potential of -60 mV that stimulated the transporter, application of GABA induces inward-directed current; during the voltage-clamp pulse to the positive potential of $+60$ mV no GABA-induced signal can be detected when current has reached steady state (see Fig. 7.14).

In the absence of GABA, the transport cycle cannot be completed; binding and unbinding of extracellular Na^+ is still possible and gives rise to slow transient current signals in response to a change in membrane potential (Fig. 7.14). As discussed before (Sect. 4.3), from the voltage dependence in the charge distribution the effective valency and the number of functional transporters can be determined.

7.2   Carriers Are Like Channels with Alternating Gates

We have now seen that carriers have at least partially characteristics of a channel, in that sense that ions cross part of the membrane through a channel-like structure (access channel), and that the actual carrier mechanism needs to occur only via a very short distance.

We can think about this like in a real channel where a boat passes the channel with locks (Fig. 7.15) where alternating gates allow a boat being transported from a lower to a higher level. What happens when a lock is broken and stays always open? Then we have a channel with only one lock gate, and only downhill passage can occur, and this indeed can happen in case of the anion exchanger. In flux measurements a tiny net flux can be detected. This can be attributed to the opening of single channels as shown in Fig. 7.16, though with an extremely low probability.

Fig. 7.15 Channel with two
lock gates as a cartoon for an
active transporter with access
channel. A boat enters the lock,
the gate closes, the boat can
leave at a higher level after the
lock is floated. If now the other
gate will also open, the boat will
flow down along the water
gradient

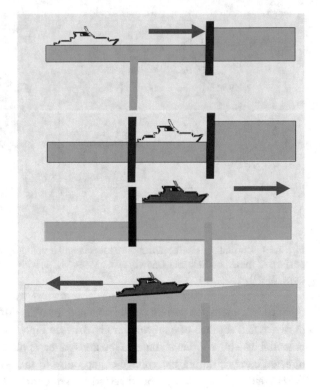

Fig. 7.16 Single-channel events from anion exchanger (adopted from Schwarz et al., 1989, Fig. 6, with kind permission from Elsevier AG, 1989)

The existence of a channel mode has also been postulated for the neurotransmitter transporters. Figure 7.17 illustrates the result of an experiment with GAT1, where the uptake of radioactively labelled neurotransmitter was measured under voltage clamp, so that the rate of uptake and current could be determined simultaneously.

Assuming that 2 Na^+ ions are transported per GABA molecule, a current can be calculated that is by order of magnitude smaller than the actually measured current (and vice versa). This suggests that in addition to the carrier mode a channel-like mode exists allowing ions to cross the membrane.

Fig. 7.17 Comparison of
GAT-mediated flux and current
at −60 mV. Grey unhatched bar
gives rate of uptake calculated
from the measured current, grey
hatched bar calculated from
measured rate of uptake

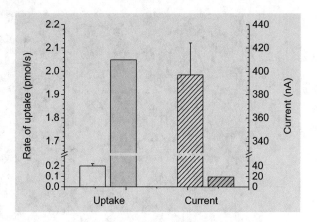

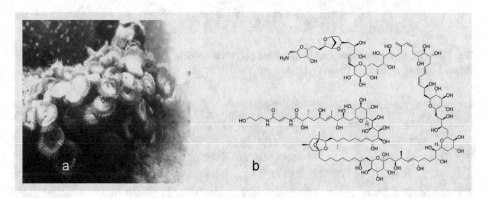

Fig. 7.18 *Palytoa* species (**a**) and structure of PTX (**b**)

For the Na⁺ pump an indication for a channel mode has not been found under physiological conditions, but can be induced by the most potent non-proteinous natural toxin, palytoxin (Fig. 7.18b), which was extracted from the coral *Palytoa*, (Fig. 7.18a).

A detailed review of pharmacological action of palytoxin has been published recently by C.H. Wu (Wu, 2014). In the presence of this toxin, the pump exhibits single-channel events (Fig. 7.19a) that allow permeation of Na⁺ and K⁺ ions. Since the pump density in a cell membrane is extremely high, such single-channel events can be seen only during the onset of toxin action. Later on, only steady-state currents (see Fig. 8.3) or elevated current fluctuations can be recorded, and fluctuation analysis can be performed (Fig. 7.19b).

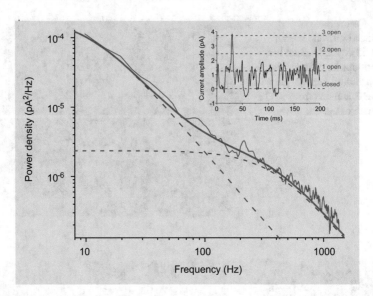

Fig. 7.19 (**a**, **b**) Power density spectrum from palytoxin-modified Na pumps (inset shows a single-channel current recording) (based on data from Vasilets et al., 2000)

Take-Home Messages

1. *Xenopus* **oocytes** can be used **as an expression system** for channel and carrier proteins. They are an ideal tool for their electrophysiological investigation.
2. **Carrier transport** can be described by reaction diagram with rates that represent transitions between different protein **conformations involving translocation** of the respective substrate.
3. In contrast, **reaction diagrams for ion channels** describe **conformation-al changes between open and closed states**.
4. **Carrier transport** can be **modulated by changes in membrane potential**
   (a) if a net charge is transported across the membrane (**electrogenic transport**)
       e.g. Na/K-ATPase, neurotransmitter transporters.
   (b) if conformational changes involve charge movements (**voltage-dependent rates**)
       e.g. anion exchanger, Na/K-ATPase.
5. Channel properties of carrier transporter can be described by "**alternating gates**".

Exercises

1. What are typical differences between channels and carriers?
2. Write down the equation for the reversal potential of the GABA transporter.

3. Which electrical signals can be detected from the Na^+,K^+-ATPase, and which information can be extracted from these signals?
4. How can current generated by channel opening be discriminated experimentally from that generated by a transporter?
5. What is the difference between primary and secondary active transport?
6. Under which circumstances exhibit transporters channel-like properties?

References

Eckstein-Ludwig, U., Fei, J., & Schwarz, W. (1999). Inhibition of uptake, steady-state currents, and transient charge movements generated by the neuronal GABA transporter by various anticonvulsant drugs. *British Journal of Pharmacology, 128,* 92–102.

Grygorczyk, R., Schwarz, W., & Passow, H. (1987). Potential dependence of the "electrically silent" anion exchange across the plasma membrane of Xenopus oocytes mediated by the band-3 protein of mouse red blood cells. *The Journal of Membrane Biology, 99,* 127–136.

Passow, H. (1986). Molecular aspects of band 3 protein-mediated anion transport across the red blood cell membrane. *Reviews of Physiology, Biochemistry and Pharmacology, 103,* 61–203.

Salonikidis, P., Kirichenko, S. N., Tatjanenko, L. V., Schwarz, W., & Vasilets, L. A. (2000). Extracellular pH modulates kinetics of the Na^+,K^+-ATPase. In K. Taniguchi (Ed.), *The sodium pump.* Elsevier Press.

Schwarz, W., Grygorczyk, R., & Hof, D. (1989). Recording single-channel currents from human red-cells. *Methods in Enzymology, 173,* 112–121.

Schwarz, W., Gu, Q., & Passow, H. (1992). Potential dependence of mouse band 3-mediated anion exchange in Xenopus oocytes. In E. Bamberg & H. Passow (Eds.), *The band 3 proteins: anion transporters, binding proteins and senecent antigens* (pp. 161–168). Elsevier Sc. Publ.

Vasilets, L. A., & Schwarz, W. (1993). Structure-function relationships of cation binding in the Na^+/K^+-ATPase. *Biochimica et Biophysica Acta, 1154,* 201–222.

Vasilets, L. A., Wu, C. H., Wachter, E., & Schwarz, W. (2000). Gating role of the N-terminus of α-subunit of the Na^+,K^+-ATPase converted into a channel by palytoxin. In Y. Suketa (Ed.), *Control and diseases of sodium transport proteins and channels.* Elsevier Press.

Wu, C. H. (2014). Pharmacological action of Palytoxin. In G. P. Rossini (Ed.), *Toxins and biological active compounds from microalgae* (Vol. 2). CRC Press Taylor & Francis Group.

Examples of Application of the Voltage-Clamp Technique

8

Contents

Abstract

This chapter shall describe with exemplary illustrations how combination of electrophysiology, molecular biology, and pharmacology can be applied to learn about structure, function, and regulation of membrane permeabilities. These characteristics form the basis of cellular function and their role in diseases.

J. Rettinger et al., *Electrophysiology*,
https://doi.org/10.1007/978-3-030-86482-8_8

In addition to the Na,K pump and the GABA transporter, as examples for active transporters, the purinergic receptor P2X will be introduced as example to illustrate the analysis of structure–function for ion channels. Electrophysiology can also be applied to understand underlying mechanisms in diseases and their treatment. This will be illustrated for the role of viral ion channels in infections and as target for antiviral drugs.

A special section illustrates how electrophysiology can be used to understand basic cellular mechanisms in traditional Chinese medicine and to investigate drug–receptor interaction in pharmacology.

Keywords

Na,K pump · GABA transporter · P2X receptor · Viral ion channels · Traditional Chinese Medicine · Pharmacology

All the techniques we have discussed, flux measurements, steady-state and transient current measurements as well as single-channel recordings, and the corresponding analysis can be applied for the analysis of structure-function relationships. Such structure–function information can be obtained if we characterise and compare the function of wild-type and chemically or genetically modified transporters by using these techniques. The latter also includes naturally occurring mutations that are the source of various diseases, an important feature to understand and cure such diseases.

For many of the transporters the amino acid sequence and the possible orientation of the protein in the membrane or even the three-dimensional structure could be determined. In the following we will illustrate the strategy of an electrophysiologist in investigating structure, function and regulation of membrane transport using as an example the Na^+, K^+-ATPase (Fig. 8.1), the neurotransmitter transporter GAT (Na^+-dependent GABA transporter (Fig. 8.4) (Sect. 8.1), and the nucleotide receptors (Fig. 8.6) that form channels in the presence of extracellular ATP (Sect. 8.2). Electrophysiology can also be applied to understand underlying mechanisms in diseases and their treatment. This will be illustrated for viral ion channels (Sect. 8.3), whose activity is essential for virus reproduction. Therefore, these channels form an ideal target for antiviral drugs.

Ancient Chinese medicine (Traditional Chinese Medicine (TCM)) became more recently also accepted in Western Medicine. At least for certain aspects a cellular basis could be demonstrated, and electrophysiology could considerably contribute (Wang et al. 2021). This shall be illustrated in Sect. 8.4. For an understanding of drug action as well as the development of new drugs for treatment of diseases, electrophysiology is a powerful method to elucidate drug–receptor interaction. As an example, this will be illustrated for essential membrane transporters (carriers) and ion channels (Sect. 8.5).

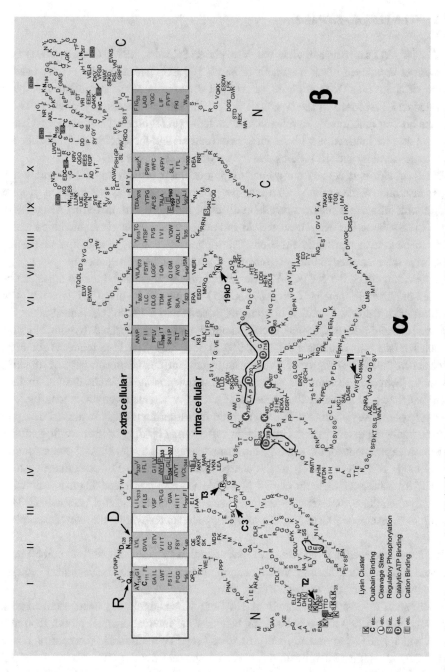

Fig. 8.1 Amino acid sequence of the Na$^+$,K$^+$ ATPase (based on Vasilets & Schwarz, 1993, with kind permission from Elsevier AG, 1993)

8.1   Structure–Function Relationships of Carrier Proteins

8.1.1   The Na$^+$,K$^+$-ATPase

The Na$^+$,K$^+$-ATPase (for a review see Vasilets & Schwarz, 1993, Glitsch, 2001) is a heterodimer composed of an α subunit of about 100 kDa and a smaller glycosylated ß subunit of about 60 kDa (see Fig. 8.1). At least 4 isoforms of the α subunit have been identified and 3 isoforms of the ß subunit that all show tissue-specific distribution. The α subunits host all functionally significant sites including ATP-binding and phosphorylation sites, the sites for interaction with the transported cations and for specific inhibitors like the cardiotonic steroids, and the binding site for palytoxin (see Fig. 7.3).

The ß subunit is necessary for proper folding of the α subunit and for assembly and proper insertion of the entire protein into the cell membrane. The combination of an α subunit with different ß subunits results in different functions and, therefore, a regulatory role of the ß subunit has been discussed. In addition, a γ subunit has been identified which seems to have also regulatory function. One option to monitor modulation of the ion transport is the measurement of pump-mediated current under voltage clamp.

In the α subunit of the Na$^+$,K$^+$-ATPase several negatively charged amino acids can be localised in intramembraneous domains (Fig. 8.1) that may be involved in interaction with the transported cations. Mutation of some of these amino acids to Ala indeed alters the apparent dielectric length of the access channel that is represented by the effective valencies z (see e.g. glutamate 334 and 960 in Fig. 8.2). The N-terminus is the area, which shows the highest degree of diversity among the different isoforms of the α subunits, and may, therefore, account for isoform-specific function in different tissues. In fact, mutations (like truncation as illustrated in Fig. 8.2) or chemical modifications by, e.g. regulatory phosphorylation (see Fig. 8.1) within the N-terminus leads to altered transport function including external cation interaction and external binding of the specific inhibitor ouabain. Measurements of the electrogenic current generated by the Na$^+$,K$^+$ pump confirm that mutation of Q_{118} and N_{129} to the charged residues R and D, respectively, causes insensitivity of the ATPase to the cardiac glycoside ouabain. It is interesting to mention that the highly flexible, cytoplasmic N-terminus interferes with the external interactions. This finding is an example illustrating the allosteric interaction within the complex protein structure.

We have described above (Sect. 7.2) that palytoxin transfers the Na$^+$,K$^+$ pump into a channel. During long-lasting voltage-clamp pulses the palytoxin-induced current inactivates at very positive potentials (Fig. 8.3a).

Inactivation at positive potentials of Na$^+$ and K$^+$ channels has been ascribed to a positively charged ball at the N-terminus blocking the internal channel mouth (ball at a chain), which leads to an inward-rectifying steady-state IV relationship (compare e.g. also Sect. 5.2.2). A similar interpretation has been suggested for the PTX-modified Na$^+$,K$^+$ pump. Indeed, the truncated mutant does not show inactivation anymore (Fig. 8.3b), while application of the truncated N-terminal peptide restores inactivation (Fig. 8.3c).

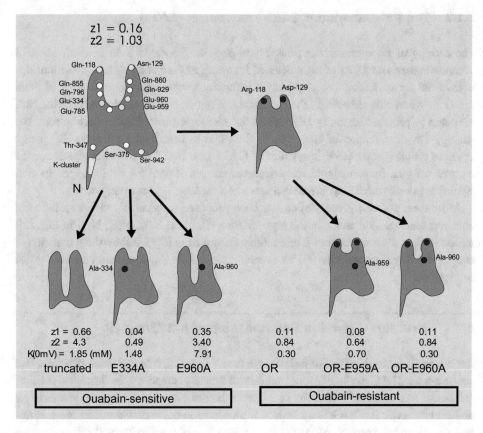

Fig. 8.2 "Dentist's" presentation of the Na^+,K^+-ATPase (based on Schwarz & Vasilets, 1996, Fig. 4, with kind permission from John Wiley and, 1996)

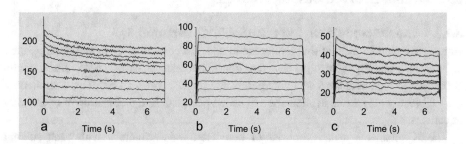

Fig. 8.3 PTX-induced current in wild-type (**a**) and truncated mutants (**b**) without N-terminal peptide, (**c**) with the peptide added to the internal solution (based on Wu et al., 2003, Fig. 4, with kind permission from Elsevier AG 2003)

8.1.2 The Na⁺-Dependent GABA Transporter (GAT1)

The activity of neurotransmitter transporters plays an important role in termination of synaptic transmission. Therefore, a detailed knowledge of structure–function relationships is essential for understanding physiology, pathophysiology, and pharmacology of brain function. We mentioned already the importance of regulatory phosphorylation of the Na^+, K^+ pump by protein kinases, and this is also the case for neurotransmitter transporters. The strategy for identification of such sites is the same as described for the Na^+,K^+ pump. Possible candidates for phosphorylation of GAT1 (see Fig. 8.4) have been mutated; the mutants are then functionally characterised and compared with the wild-type transporter. Again transporter-mediated current can serve as a measure for transport activity.

More recently, also glycosylation has been proposed to be involved in regulation of transport function. Mutations in the glycosylation sites (see N_{176}, N_{181}, N_{184} in Fig. 8.4) indeed leads to altered transport. The example illustrated in Fig. 8.5 shows that mutation of two of the three asparagines (N) to aspartic acid (D) leads to reduced sensitivity for extracellular Na^+.

8.2 Structure–Function Relationships of Ion Channels

Ion channels are classified due to their diverse modes of function or molecular structure. This can be ion selectivity, modes of gating, numbers of subunits comprising the functional channel, number of transmembrane domains or sequence homologies. Classically, ion channels are divided into major families having important properties in common like gating mechanisms and ion selectivity. Some of the families will briefly be presented in the next paragraphs.

8.2.1 Families of Various Ion Channels

8.2.1.1 The Voltage-Gated Ion Channel (VIC) Superfamily

Although some members of this family are in addition controlled by ligand binding, their activity is generally controlled by the transmembrane electrical field. Functionally characterised members are selectively permeable for K^+, Na^+, or Ca^{2+} ions. Members of the VIC family play a role in generation of action potentials and modulation of excitability of cells.

The K^+ channels usually consist of homotetrameric structures with each subunit possessing six transmembrane domains. At least ten types of K^+ channels are known, each responding in different ways to different stimuli: voltage-sensitive (Ka, Kv, Kvr, Kvs, and Ksr), Ca^{2+}-sensitive (BKCa, IKCa, and SKCa), and receptor-coupled channels (KM and KACh). There are also tetrameric channels in which each subunit possesses two transmembrane (TM) domains that are homologous to TM-domain 5 and 6 of the six TM-domain type (inward rectifier Kir).

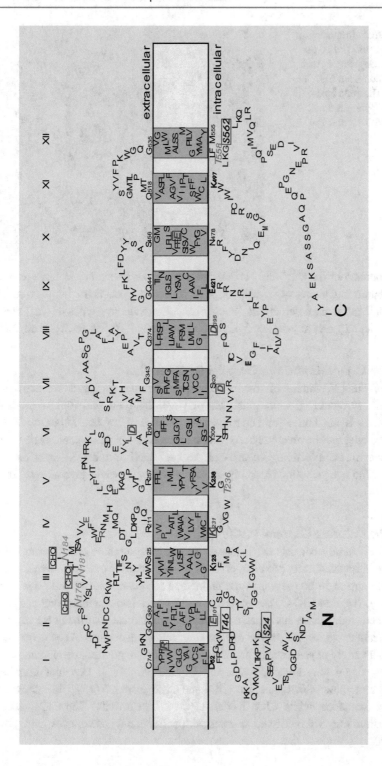

Fig. 8.4 Amino acid sequence of mouse GAT1

Fig. 8.5 Effect of mutation of glycosylation sites (based on (Liu et al., 1998), Fig. 5, with kind permission from Elsevier AG, 1998). The code letters for amino acids represent the positions 176, 181, and 184

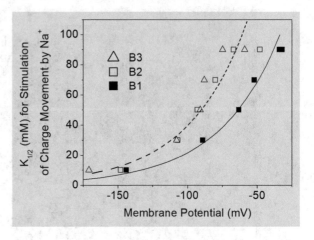

The α subunits of the Ca^{2+} and Na^+ channels are about four times as large as the K^+-channel subunits and possess 4 sequence repeats with each repeat being homologous to the single subunit in the homotetrameric K^+ channels. There are five types of Ca^{2+} channels (L, N, P, Q, and T), and at least six types of Na^+ channels (I, II, III, µ1, H1, and PN3).

8.2.1.2 The Ligand-Gated Ion Channel (LIC) Family

Members of the LIC family of ionotropic neurotransmitter receptors are activated by acetylcholine, serotonin, glycine, glutamate or γ-aminobutyric acid (GABA). All these receptor channels are homo- or heteromers of three, four, or five subunits. The best characterised ones are the nicotinic acetylcholine receptors which are pentameric channels of $\alpha_2\beta\gamma\delta$ subunit composition. Channels of the LIC family are selective for cations or anions (e.g., the acetylcholine receptors are cation-selective while glycine receptors are anion-selective).

8.2.1.3 The Chloride Channel (ClC) Family

The large ClC family consists of dozens of sequenced proteins derived from bacteria, plants, and animals. These proteins exhibit 10–12 putative transmembrane α-helical domains and appear to be present in the membrane as homodimers. While one member of the family, *Torpedo* ClC-O, has been reported to have two channels, one per subunit, others are believed to have just one. All functionally characterised members of the ClC family are permeable for Cl^-, some of them are voltage-dependent. These channels are involved in a variety physiological functions (cell volume regulation, membrane potential stabilisation, signal transduction, transepithelial transport, etc.). Different homologues exhibit different anion selectivities, i.e., ClC4 and ClC5 share a $NO_3^- > Cl^- > Br^- > I^-$ conductance sequence, while ClC3 has an $I^- > Cl^-$ selectivity. The ClC4 and ClC5 channels exhibit outward rectifying currents with currents only at voltages more positive than +20 mV.

8.2.1.4 The Gap Junction-Forming (Connexin) Family

Gap junctions consist of clusters of closely packed pairs of transmembrane channels, the connexons, through which small molecules diffuse between neighbouring cells. The connexons consist of homo- or heterohexameric arrays of connexins, and the connexon in one plasma membrane docks end-to-end with a connexon in the membrane of a closely opposed cell. Over 15 connexin subunit isoforms are known. They vary in size between about 25 kDa and 60 kDa. They have four putative transmembrane α-helical domains. A dodecameric channel is formed by two hexamers, and therefore, consists of 48 transmembrane domains in total.

8.2.1.5 The Epithelial Na$^+$ Channel (ENaC) Family

The ENaC family consists of more than 20 sequenced proteins from animals exclusively. There exist voltage-insensitive ENaC homologues in the brain. Some of these proteins are involved in touch sensitivity, other members of the ENaC family, the acid-sensing Na$^+$ channels (H$^+$-gated), ASIC1–3, mediate pain sensation in response to tissue acidosis. The (FMRF-amide)-activated Na$^+$ channel is the first peptide neurotransmitter-gated ionotropic receptor that was sequenced. All members of this family exhibit a topology with intracellular N- and C-termini, two transmembrane spanning segments, and a large extracellular loop. Three homologous ENaC subunits, α, β, and γ, have been shown to assemble to form the highly Na$^+$-selective channel with the stoichiometry $\alpha\beta\gamma$ in a heterotrimeric architecture.

8.2.1.6 Mechanosensitive Ion Channels

Another interesting group of ion channels are those that are gated by mechanical stress allowing transduction of mechanical forces into electrical signals (see e.g. Delmas & Coste, 2013; Ranade et al., 2015). In particular, the patch-clamp technique with applying mechanical or osmotic stress to a cell and recording electrical responses reveals a continuously growing number of mechanosensitive or stretch-activated (SAC) ion channels (Nilius & Honoré, 2012). A class of mechanosensitive channels that was conserved throughout evolution is formed by the Piezo channels (Bagriantsev et al., 2014). Another large class of ion channels also sensitive to physical stimuli is formed by the transient receptor potential (TRP) channels (Christensen & Corey, 2007).

The mechanosensitive channels are involved in all kinds of mechanosensation including, e.g. touch sensation, hearing or regulation of tone of smooth muscle fibres. The transfer from an open to a closed configuration can be governed by mechanical forces on the membrane protein via the lipid bilayer or via tethered cytoskeletal or extracellular structures.

8.2.2   ATP-Gated Cation Channel (ACC) Family

In the following we will present the strategy how electrophysiology can be utilised to learn about structure–function relationships of ion channels. As example we will use the ATP-gated ion channels and focus on members of the P2X receptor family.

8.2.2.1 Structure and Classification of P2X Receptors

The class of ATP-gated ion channels which belongs to the group of ligand-gated ion channels will be described in more detail since we will use members of this group as another example for application of electrophysiological techniques.

Because the common feature of these channels is the sensitivity for extracellular ATP, they are termed nucleotide receptors. Another commonly used name is P2X receptors, and formerly they were known under the name purinoceptors. Although it was already proposed in 1972 (Burnstock, 1972) that ATP plays a functional role as a neurotransmitter, it took until 1994 (Brake et al., 1994; Valera et al., 1994) that the first two isoforms $P2X_1$ and $P2X_2$ could be cloned from rat vas deferens and PC12 cells, respectively.

From hydropathy analysis a now commonly accepted secondary structure with intracellular N- and C-termini, a large extracellular loop and two transmembrane domains could be derived (see Fig. 8.6) which resemble the architecture of the ENaC family. As also known from other ion channel proteins, the functional channel unit is composed of more than one subunit. A quaternary structure of three subunits forming the functional receptor was published (Nicke et al., 1998), a stoichiometry different from what is known from other ligand-gated ion channels that are usually composed from tetramers or pentamers.

Nowadays seven isoforms (Table 8.1) and various splice variants are known, which have more or less different pharmacological and functional properties (Burnstock, 1999). Since all the seven isoforms are ionotropic ion channels (meaning that receptor and channel function are localised on the same protein multimer) electrophysiology is the method of choice to characterise their functional properties.

Before we want to introduce some of the electrophysiological data obtained for the $P2X_1$ and $P2X_2$ receptor, we want to give a short overview of the most evident differences between the subtypes. Although extracellular ATP is able to activate each of the ATP receptors, the ATP derivative α-β-methylene-ATP is an agonist only for $P2X_1$ and $P2X_3$. More or less pronounced desensitisation is obtained for all P2X receptors subtypes, which means that the agonist induces an opening of the ion channels but then channels close even in the presence of the agonist.

Fast inactivation within about a second has been found for the $P2X_1$, $P2X_3$ and slightly slower for the $P2X_4$ receptors, whereas the other subtypes desensitise only slowly and incompletely. A special feature of the $P2X_7$ receptor is the formation of a large pore with 3 to 5-nm diameter after prolonged application of agonist. A similar behaviour, where at least the ion selectivity changes after prolonged activation, is also discussed for the other subtypes.

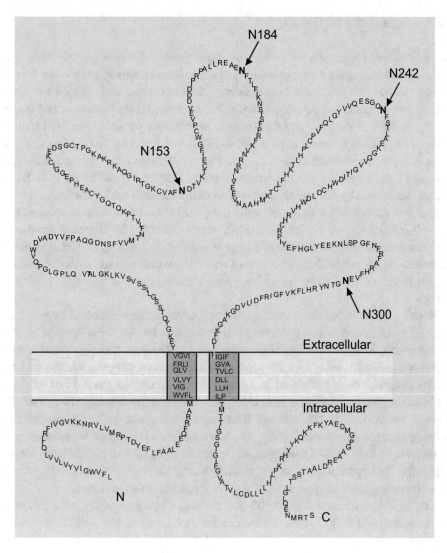

Fig. 8.6 Amino acid sequence of human P2X$_1$

Table 8.1 Comparison of the known P2X subtypes

	P2X$_1$	P2X$_2$	P2X$_3$	P2X$_4$	P2X$_5$	P2X$_6$	P2X$_7$
α-β-methylene-ATP	+	−	+	−	−	−	−
Desensitisation	Fast & complete	Slow & incompl.	Fast & complete	Middle & incompl.	Slow & incompl.	Slow & incompl.	Slow & incompl.

8.2.3 Experimental Results

In the following paragraphs we will present some results on functional properties of the P2X$_1$ and P2X$_2$ receptor. Although analysis of single-channel activity can provide the most valuable insight into ion-channel characteristics, it is sometimes difficult to detect single-channel events depending on single-channel conductance, gating behaviour, and channel density. In case of the P2X receptors, mainly macroscopic currents have been analysed either due to the fast desensitisation (P2X$_1$ and P2X$_3$) or due to the fast gating (flickering) of the P2X$_2$ receptors (Ding & Sachs, 1999). Figure 8.7 shows a current trace of the P2X$_1$ receptor in an outside-out oocyte-membrane patch. Only at the end of the trace current fluctuations caused by the opening and closing of the channels can be resolved.

The data shown in the following were recorded by the two-electrode voltage-clamp (TEVC) technique (Sect. 4.1.3) with *Xenopus* oocytes (Sect. 7.1.2). Therefore, the currents represent the simultaneous activity of a huge number of ion channels (millions) leading to currents in the µA-range. *Xenopus* oocytes do not possess endogenous nucleotide receptors and are, therefore, an ideal tool for studying these channels after injection of the appropriate cRNAs.

Since P2X$_1$ and P2X$_3$ receptors show fast activation and desensitisation (within 1 s), a fast and well-defined solution exchange is a prerequisite for correctly time-resolved measurements. Due to the large size of the oocyte (diameter 1.2 mm) a fast solution exchange is not easily to achieve, but is possible with a special design: a solution exchange system where the oocyte is placed in a small oocyte chamber of about 10 µl volume in combination with fast perfusion (200 µl/s). This design allows exchange time (from 5 to 95%) in the range of a 100–200 ms. This can be determined by activating expressed nicotinic acetylcholine receptors with 30 µM acetylcholine (ACh) or by monitoring the current change induced by a concentration change from Na$^+$-rich to Na$^+$-deficient solution (Fig. 8.8). The reproducibility of the solution change is obvious in Fig. 8.8a where 7 consecutive responses to application of ACh at 1 min intervals are shown.

Although the exchange time of 100–200 ms is fast in regard of the large size of the oocyte, one should keep in mind that the currents that are elicited by a solution exchange

Fig. 8.7 Current trace of the P2X$_1$ receptor in an outside-out oocyte-membrane patch at a holding potential of −60 mV with superimposition of several single-channel events (Jürgen Rettinger, unpublished)

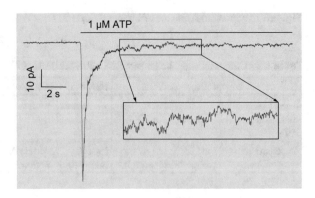

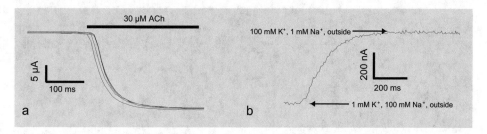

Fig. 8.8 Time course of solution exchange with nAChR-mediated currents (**a**) and with endogenous Na^+ conductance of the oocytes (**b**) (Jürgen Rettinger, unpublished)

(for example, ATP application to P2X receptors) are greatly influenced by the concentration change that occurs during the first 200 ms. A significantly faster solution change on intact oocytes will be hard to get, and if this becomes necessary, one has to use the cell-free patch-clamp method (see Sect. 4.4.1 and 4.4.2), which allows for exchange times even in the sub-millisecond range.

8.2.3.1 The P2X₁ Receptor

As already mentioned, $P2X_1$ is a fast desensitising receptor that opens an intrinsic ion channel non-selectively permeable for small cations when challenged with extracellular ATP. Figure 8.9 shows the current in response to 1 μM ATP for the first application of ATP (left peak) and for consecutive ATP applications at 5-min intervals and a holding potential of −60 mV.

The receptor desensitises in the presence of ATP completely within about 1 s. After a period of 5 minutes in ATP-free solution typically only about 25% of the initial receptor current can be restored. This means that desensitisation is fast but the recovery from the desensitised state is slow. As a matter of fact, tens of minutes are necessary to restore the full initial current response. The response of the $P2X_1$ receptors is dependent on the extracellular ATP concentration with an EC_{50} value (the concentration that is needed for half-maximum response) of about 1 μM.

Figure 8.10a shows a typical experiment for determination of the EC_{50} value by activating the receptors with different ATP concentrations ranging from 0.03 μM to 30 μM. Figure 8.10b shows the dose–response curve of the receptor for ATP.

Fig. 8.9 $P2X_1$ receptor currents activated with 1 μM ATP at 5 min intervals at a holding potential of −60 mV (Jürgen Rettinger, unpublished)

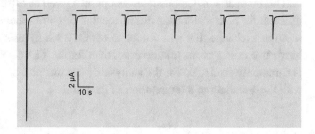

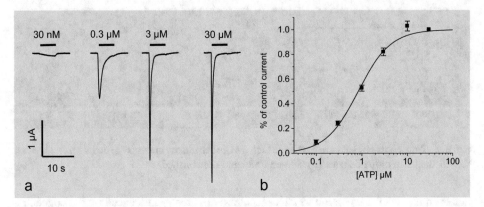

Fig. 8.10 P2X$_1$ receptor currents activated with different ATP concentrations (**a**) and the ATP dose–response curve (**b**) (EC$_{50}$ = 0.6 μM). Receptors were expressed in *Xenopus* oocytes after RNA-injection

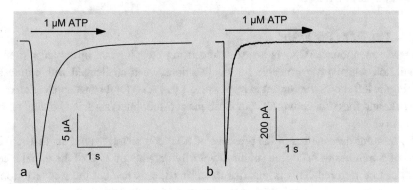

Fig. 8.11 Comparison of P2X$_1$-generated current activated by 1 μM ATP in TEVC (**a**) and an outside-out macro-patch (**b**) at a holding potential of −60 mV (Jürgen Rettinger, unpublished)

Since the P2X$_1$ receptor shows fast activation after application of extracellular ATP, it is interesting to compare the signals that can be measured with the TEVC and the patch-clamp method. Figure 8.11a shows the current trace from an intact oocyte and Fig. 8.11b that from an excised outside-out patch. In both cases the channels are activated by 1 μM ATP.

This comparison demonstrates that onset as well as offset of the current are significantly different for both methods, a difference that can be explained by the difference in speed of solution exchange that was complete within 2 ms by using the patch-clamp method, and hence two orders of magnitude faster than for the TEVC. Nevertheless, the TEVC method is commonly used, also for the analysis of fast receptor currents, but one should be careful with the quantitative interpretation of the data.

8.2.3.2 The P2X$_2$ Receptor

As an example for a nearly non-desensitising P2X receptor we want to introduce to you the P2X$_2$ receptor which, like all members of the P2X family, opens an intrinsic ion channel after extracellular application of an appropriate agonist. Analogous to the P2X$_1$ receptor, Fig. 8.12 shows P2X$_2$ currents activated by different ATP concentrations (left) and the corresponding dose–response curve (right) with an EC$_{50}$ value of about 30 µM.

8.2.3.3 Effect of Glycosylation on P2X$_1$ Receptor Function

After the presentation of these pharmacological data, we will concentrate in the following on the use of electrophysiology in combination with biochemical methods. This combination can give answers to questions concerning the functional consequences of structural changes in the channel protein.

Analysis of the amino acid sequence of the P2X$_1$ receptor revealed the presence of 5 consensus sequences for putative glycosylation localised on the extracellular loop of the protein (N1-N5, see Fig. 8.6). Biochemical analysis of the receptor protein (expressed in oocytes) could demonstrate that four out of these five sites are used for glycosylation. A question that cannot be answered with biochemical methods is whether functional properties of the expressed receptors (ATP dependence, kinetic parameters, etc.) are changed. Therefore, different constructs at which one or several glycosylation sites were removed by mutation of single amino acids were characterised electrophysiologically in respect to its ATP dependency and magnitude of ATP-activated current.

The electrophysiological analysis could show that removal of the third N- glycosylation site alters the apparent affinity of the receptor for ATP by a factor of about 3. All the other N-glycosylation site do not seem to have any influence on the ATP sensitivity. Figure 8.13 gives a graphical representation of these results.

Taken together, results from the combination of molecular biology, biochemistry, and electrophysiology could demonstrate that an increased number of missing glycosylation

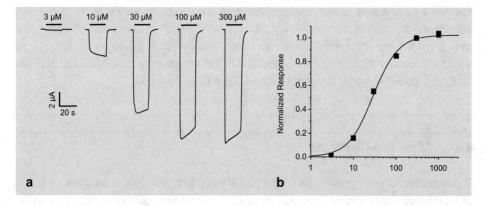

Fig. 8.12 P2X$_2$ receptor currents activated by different ATP concentrations (**a**) and the ATP dose–response curve (**b**) EC$_{50}$ = 29 µM, HP = −60 mV) (Jürgen Rettinger, unpublished)

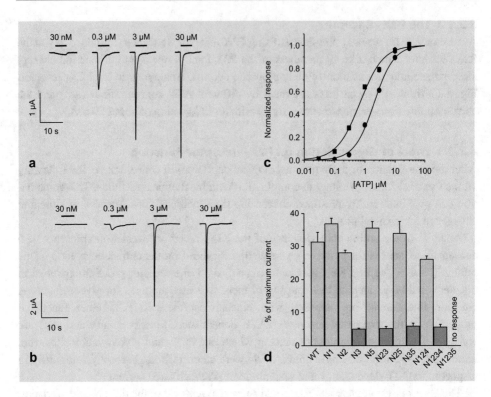

Fig. 8.13 P2X1 receptor currents of wild-type (**a**) and N3-mutated (**b**) subunits activated by different ATP concentrations and ATP dose–response curves for wild-type and mutant ΔN3, respectively (**c**). The bar graph (**d**) shows the amount of current that was measured at 0.3 μM relative to the maximal current at 30 μM; it demonstrates that all mutants lacking the N-glycan at position N3 exhibit lower affinity for ATP (according to Rettinger et al., 2000, Fig. 4, Copyright 2000, based on Creative Commons CC by Elsevier)

sites led to a decrease in surface appearance of the receptors (reflected by decreasing currents and the decreasing intensity of bands on the SDS-gel). A functional difference was only found for the constructs that were deficient of the third glycosylation site. These constructs showed the effect of a decreased ATP sensitivity with an EC_{50} value of 2 μM compared to the wild-type receptor with an EC_{50} value of 0.6 μM.

8.3   Viral Ion Channels

In this section we will present the strategy how electrophysiology can be combined with pharmacology (but see also Sect. 8.5) to develop new drugs against viral infections. As example we will use the viral ion channels as a target for antiviral drugs.

Table 8.2 Examples of viruses and their viral ion channels that are formed by multi-homomers, each subunit consisting of 1–3 transmembrane segments (TMS). Channels printed in bold are dealt with in this Sect. 8.3

Virus family	Virus	Ion channel	Characteristics	Functional units
Coronaviridae	SARS-CoV	**3a**	≈30 pS[a] Monovalent cation	Tetramer[b] (3 TMS)
	SARS-CoV-2	**3a**	Cation channel	Tetramer[c] (3 TMS)
		E	Ion channel[d]	Pentamer[e] (1TMS)
Orthomyxoviridae	Influenza A (swine flu)	**M2**	< fS Proton	Tetramer[f] (1 TMS)
	Influenza B	BM2	< fS Proton	Tetramer[g] (1 TMS)
Picornaviridae	Poliovirus	2B	Non-selective	Tetramer[h] (2 TMS)
Retroviridae	HIV-1	**Vpu**	≈20 pS Monovalent cation	Pentamer[i] (1 TMS)
Flaviviridae	HCV (hepatitis C)	p7	20–100 pS Monovalent cation	Hexamer[j] (2 TMS)

[a]Sakaguchi et al. (1997)
[b]Lu et al. (2006)
[c]Kern et al. (2021)
[d]Ruch & Machamer (2012)
[e]Mandala et al. (2020)
[f]Schwarz et al. (2012)
[g]Mould et al. (2003)
[h]Patargias et al. (2009)
[i]Cordes et al. (2001)
[j]Luik et al. (2009)

The genomes of various viruses encode for proteins that may form ion-selective channels in the infected cell. These ion channels play important roles in the viral life cycle, and therefore, may represent a target for new antiviral drugs.

Table 8.2 lists examples of such channels (see also Wang et al., 2011; Krüger & Fischer, 2009; Fischer & Sansom, 2002). The pores are formed by multi-homomers.

The viral life cycle involves a sequence of steps that actually all may form a target for antiviral drugs. As an example, Fig. 8.14 illustrates various steps in the life cycle of coronavirus.

The virus attaches to the host cell, followed by incorporation of the virus, uncoating of the viral genome and replication processes with transcription and translation; finally, new viral particles are assembled, and the viruses are released from the host cell to infect new cells. In case of coronavirus, the release is dependent on the activity of an ion channels,

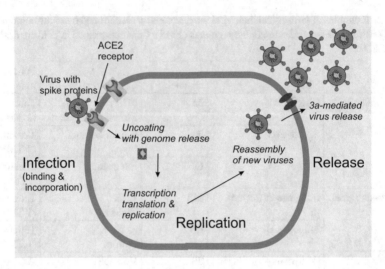

Fig. 8.14 Viral life cycle of SARS-CoV. Binding of viral spike protein with angiotensine-converting-enzyme-2 (ACE2) receptor of the host cell is followed by virus uptake, uncoating, and (transcription and) translation of the viral genome. After assembly of new viruses, 3a channel activity stimulates exocytotic release of the viruses from the host cell (see also Schwarz et al., 2012)

which is encoded by the viral genome and inserted into the membrane of the host cell. For SARS coronavirus (SARS CoV) the membrane protein is called 3a protein. Inhibition of any of these steps could actually represent a potential target for antiviral drugs.

In the following we will illustrate how inhibition of ion-channel function can interfere with the viral life cycle. In Table 8.2, the respective viral ion channels dealt with in this section are printed in bold.

8.3.1   The 3a Protein of SARS Coronavirus

For the 3a protein of SARS coronavirus (comp. Figure 8.14) it was demonstrated (Lu et al., 2006) that tetramers form the ion channel (Table 8.2) that is incorporated into the membrane of the infected cell (see Fig. 8.15a).

Patch-clamp analysis (Fig. 8.16) revealed single-channel conductance of about 20–30 pS. The channel is selectively permeable for monovalent cations with highest selectivity for K^+. As a result of 3a channel activity the membrane potential will depolarise, which results in Ca^{2+} channel activation. The increase in intracellular Ca^{2+} activity then facilitates the exocytotic release of viruses from the host cell (Lu et al., 2006).

Voltage clamp is an easy method to detect and analyse function of the viral ion channels. Using again the model system *Xenopus* oocyte as an expression system (compare 7.1.2), drugs can be screened with respect to their interaction with the 3a protein. As a general rule,

Fig. 8.15 Fluorescence-
labelled 3a protein becomes
visible in infected Vero E6 cells
(based on Lu et al., 2006) (**a**).
Orientation of 3a monomer in
the cell membrane (**b**) (based on
Lu et al., 2006, Copyright, 2006,
National Academy of Sciences,
USA)

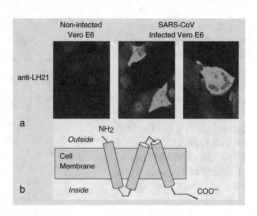

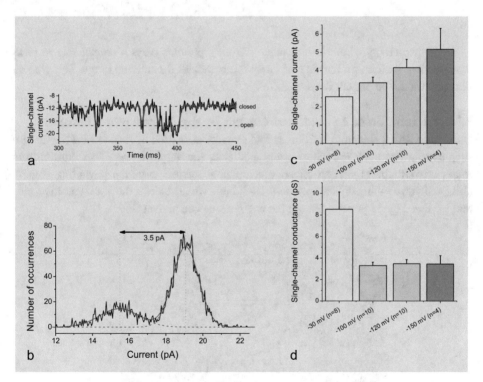

Fig. 8.16 Single-channel recordings from 3a protein expressed in *Xenopus* oocytes at -150 mV (**a**),
histograms of current at 100 mV (**b**), and voltage dependence of the single-channel current (**c**) and
conductance (**d**) (based on unpublished data by Wei Lu and Wolfgang Schwarz)

a method needs to be elaborated that allows extracting the current component of interest
from total membrane current. Injection of cRNA for the 3a protein of SARS-CoV results in
elevation of membrane current that can be blocked by 10 mM Ba^{2+} in the extracellular
medium (Fig. 8.17).

Fig. 8.17 Open triangles
represent current–voltage
dependency in oocytes not
expressing 3a protein, open and
filled squares those in 3a-
expressing *Xenopus* oocytes in
the absence and presence of
10 mM Ba^{2+}, respectively (based
on data from Lu et al., 2006,
Copyright 2006 National
Academy of Sciences, USA)

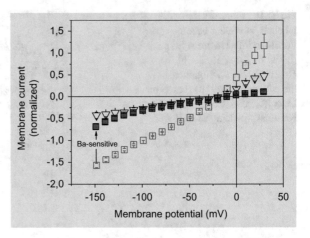

Since the activity of these ion-channel-forming proteins plays a crucial role in virus
reproduction, inhibitor of the channels may be candidates for antiviral drugs. This we will
illustrate for the 3a protein for SARS-CoV-1.

8.3.1.1 Inhibition of 3a-Mediated Current by the Anthrachinon Emodin

During the SARS epidemic in 2003, herbal extracts were used in Asia to treat the disease
supplementary to treatment with western medicine. Among those were also extracts from
Rhei radix (Rhubarb), and an effective component seemed to be the phytodrug emodin
(1,3,8-trihydroxy-6-methylanthracene-9,10-dione), which blocks the Ba^{2+}-sensitive current mediated by 3a protein, but not the endogenous component (Fig. 8.18).

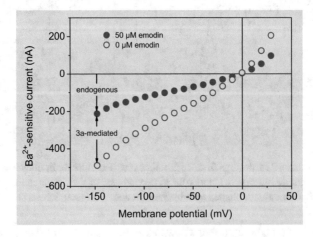

Fig. 8.18 Voltage dependency of Ba^{2+}-sensitive current in oocytes expressing 3a protein in the
absence (open circles) and presence (filled circles) of 50 μM emodin, which completely inhibits 3a-
mediated current, but leaves endogenous component unaffected (based on Schwarz et al., 2011, Fig. 1
and 2, with kind permission from Elsevier AG 2011)

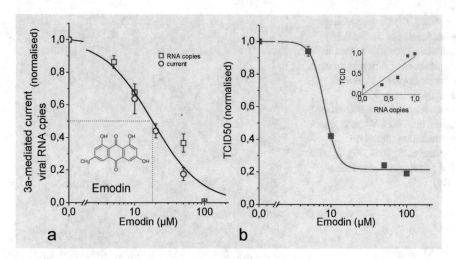

Fig. 8.19 (**a**) Emodin-dependent inhibition of 3a-mediated current and of extracellular viral cRNA, and (**b**) emodin-dependent titre concentration (inset illustrates correlation of titer concentration and number of cRNA copies (based on Schwarz et al., 2011, Fig. 4, with kind permission from Elsevier AG 2011)

Emodin not only inhibits 3a-mediated current, but with the same IC_{50} value of about 20 μM the number of viral RNA copies in the medium of infected cells (Fig. 8.19a). The number of RNA copies correlates with titre (Fig. 8.19b) indicating that the RNA originates from intact viruses. This proves that emodin can act as an antiviral drug and may form the basis for the development of new drugs against coronavirus infection.

8.3.1.2 Inhibition of 3a-Mediated Current by the Kaempferol Glycoside Juglanin

Other effective phytodrugs are the flavonoids, and especially kaempferol glycosides are highly potent inhibitors of viral ion channels (Schwarz et al., 2014).

The kaempferol glycoside juglanin effectively blocks 3a-mediated current (Fig. 8.20a), and it is by even an order of magnitude more effective ($IC_{50} \approx 2$ μM) than emodin (Fig. 8.19a).

8.3.2 Channel Proteins of SARS Coronavirus-2

The SARS coronavirus-2 (CoV-2), which is responsible for the recent worldwide pandemic CoViD-19, codes in its RNA genome also for 3a protein. This SARS-CoV-2 3a protein, like 3a of SARS-CoV-1, forms as a homotetramer (see Table 8.2) an ion channel permeable for monovalent cations (Jonas Friard, John Hanrahan, Silvia Schwarz, Wolfgang Schwarz, unpublished). In oocytes of *Xenopus laevis* with exogenously expressed 3a protein of

Fig. 8.20 (a) Current–voltage
dependencies of 3a-mediated
current in *Xenopus* oocytes in
the presence of different amount
of the kaempferol glycoside
juglanin (inset in **b**). (**b**)
Dependence of 3a-mediated
current (at −60 mV) on juglanin
concentration (based on
Schwarz et al., 2014, Fig. 3, with
kind permission from Thieme,
2014)

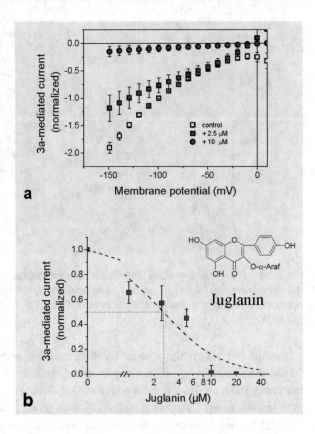

SARS-CoV-2, activity of this channel can, like 3a of CoV-1, be detected as Ba^{2+}-sensitive
currents (Fig. 8.21a).

Also the envelope protein E of SARS coronavirus forms as homomultimer (Table 8.2);
five one-TMS subunits form the porous structure (see Fig. 8.21b) (Mandala et al., 2020). It
has been suggested that its activity as an ion channel plays an essential role in the release of
the virion from the infected cell (for a review see Ruch & Machamer, 2012). Similar to the
3a protein, the E protein seems to be permeable for monovalent cations. Expression of E
protein of SARS-CoV-2 in *Xenopus* oocytes results in a Ba^{2+}-sensitive cation channel (Fig.
8.21) permeable for Na^+ and K^+ (Jonas Friard, John Hanrahan, Silvia Schwarz and
Wolfgang Schwarz, unpublished).

8.3.3   The Viral Protein Unit (Vpu) of HIV-1

The viral protein unit Vpu of HIV-1 is also a membrane protein but with only one
transmembrane segment (Fig. 8.22a), and a channel permeable for monovalent cations is
formed by a pentamer (see Table 8.2).

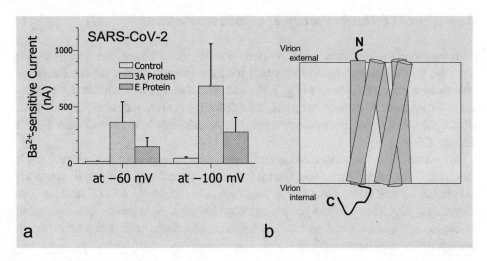

Fig. 8.21 (a) Ba^{2+}-sensitive current at -60 and -100 mV mediated by E protein expressed in Xenopus oocytes. (b) Orientation of the TMS of E protein of SARS-CoV-2 and formation of a pore by pentamer. (Jonas Friard, John Hanrahan, Silvia Schwarz and Wolfgang Schwarz, unpublished). Data are means + SEM ($n = 4$–8)

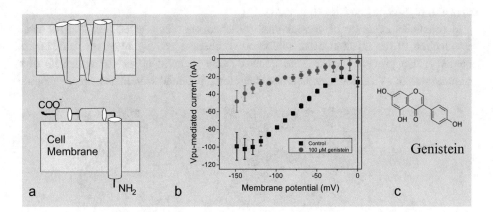

Fig. 8.22 (a) Orientation of Vpu monomer in the membrane (bottom) and formation of ion channel by pentamer (top). (b) Inhibition of Vpu-mediated current by 100 μM of the flavonoid genistein (c) (based on Sauter et al., 2014, Fig. 7, with kind permission from Thieme, 2014)

The activation of the channel is necessary for virus release (Schubert et al., 1996), and the search for Vpu channel blockers as antiviral drug has been suggested. Vpu-mediated current can also be monitored as Ba^{2+}-sensitive current (Fig. 8.22b), and can partially be block by the flavonoid genistein.

8.3.4   The M2 (Matrix Protein 2) of Influenza a Virus

Also the genome of influenza A virus (swine flu virus) encodes for an ion channel. As for the Vpu, this M2 protein has only a single transmembrane segment; but the channel is formed by a tetrameric structure (Fig. 8.23a and see Table 8.2) that exhibits low conductance for protons. With decreasing pH, the M2-mediated current dramatically increases (Fig. 8.23b). Activity of the channel also plays an essential role in virus production (see De Clercq, 2006).

In contrast to the 3a protein, M2 is an integral membrane protein of the virion and is involved in the uncoating of the viruses by permitting passage of protons across the membrane of the viron. One of the most effective inhibitors of M2 function was amantidine, a potent medicine for treatment of influenza A infection. Meanwhile, all influenza A viruses became amantadine-resistant, and world-wide laboratories search intensively for substitutes.

8.3.4.1 Inhibition of M2-Mediated Current by Kaempferol Triglycoside

Kaempferol glycosides may also form the basis for new drugs against influenza A. A triglycoside was found that effectively inhibits pH-sensitive current when M2 was expressed in *Xenopus* oocytes (Fig. 8.24).

In conclusion, activity of various viral ion channels seems to be essential for virus reproduction in the infected cells. Inhibition of channel activity by drugs will, hence, counteract that process allowing the infected body to build up or strengthen its own immune system. Viral ion channels are, therefore, potential candidates for developing

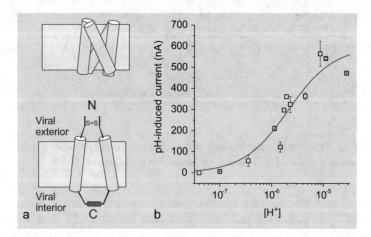

Fig. 8.23 (**a**) Illustration of structure of the M2 protein of influenza A virus: lower part shows two monomers that interact with their N and C termini, upper part shows the formation of the tetrameric channel. (**b**) pH dependency of current mediated by M2 protein expressed in *Xenopus* oocytes (Silvia Schwarz and Wolfgang Schwarz, unpublished)

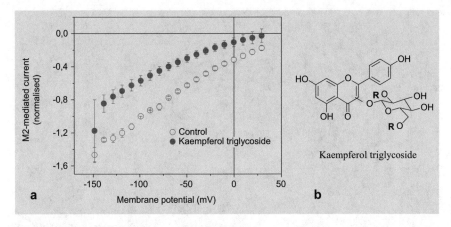

Fig. 8.24 Inhibition of M2-mediated current (**a**) by 20 µM kaempferol triglycoside (**b**) (Silvia Schwarz and Wolfgang Schwarz, unpublished)

new antiviral drugs, and voltage clamp is an easy method to screen drugs, and to detect and analyse inhibition of the ion channels. Screening a large number of natural drugs revealed that kaempferol derivatives and anthrachinones may be interesting candidates.

8.4   Electrophysiology as a Tool in Chinese Medicine Research

In Traditional Chinese Medicine (TCM) treatment of diseases is based on application of acupuncture/moxibustion and Chinese herbal formulae. Stimulation of organ-specific acupuncture points (acupoints), located close to the body surface, leads to signals that seem to spread along a network of conduits beneath the so-called meridians to the affected sites; at least for some signals, nerves seem to form the structure underlying the meridians (comp. Figure 8.25). TCM treatment intends to interfere with all these structures cartooned in Fig. 8.25, and it is based on the concepts of ying/yang (阴/阳) and qi sensation (气). In contrast, Western Medicine is based on understanding physiological and biochemical processes in the occurrence and treatment of pathophysiological conditions. Since the treatment of various selected diseases by TCM, in particular pain, has been accepted to be effective also by Western Medicine, there must exist physiological and biochemical underlying mechanisms, and it is, therefore, an interesting task to explore the basis.

Treatment of pain (see e.g. Gereau et al., 2014) and cardio-vascular diseases (see e.g. Hao et al., 2017) by TCM is well established though the underlying mechanisms are to a large extent incompletely understood. Here we like to illustrate the usefulness of electro-physiological methods in the investigation of the mechanisms of TCM with special focus on analgesic effects.

Cell electrophysiological research is based to a large extent on voltage-clamp methods (Sect. 3.4.5). The technique allows to analyse electrical signals originating from charge

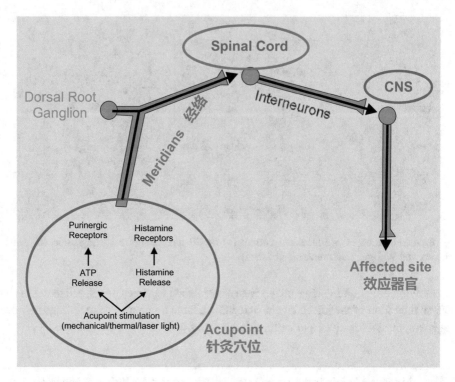

Fig. 8.25 Physiological basis of acupoint stimulation, signal transduction, and final effect. Acupoint stimulation leads to release of mediators, which stimulate sensory nerve fibres that originate from dorsal root ganglion modulating via the CNS physiological function in the affect site

movement across or within the cell membrane mediated by the respective membrane proteins. These signals provide detailed information about physiological and pathophysiological function of the protein. Therefore, information about interaction of proteins with Chinese medical stimuli as well as Chinese herbal drugs can be extracted from electrophysiological experiments. The power of electrophysiology is further gained by combination with fluorescence measurement, molecular biological methods including protein expression as well as pharmacology (comp. Sect. 8.5). With selected examples we like to illustrate the usefulness of electrophysiology in TCM research.

From the point of view of Western medicine, the cellular counterparts of acupoints, conduits, and affected sites as well as underlying physiological and biochemical processes are of interest. In previous work our laboratory focused on exploration of cellular events occurring in acupoints (Zhang et al., 2008; Wang et al., 2010) as well as in affected sites (Pu et al., 2012; Xia et al., 2006; Deng et al., 2009). Recent investigations (Burnstock, 2009; Goldman et al., 2010; Wang et al., 2013; Shen et al. 2019, 2021) support that release of ATP and the metabolic products play a key role in acupoint responses to acupuncture/moxibustion.

Since analgesia is a well-accepted effect of physical stimulation of acupuncture points, we can learn about the cellular mechanisms experimentally by following up changes is pain threshold in response to the physical stimuli.

8.4.1   Mechanisms in Acupuncture Points

8.4.1.1 Mast-Cell Degranulation

The treatment of acupoints is the application of physical stimuli in the form of mechanical stress (needling acupuncture or acupressure (tuina)), as well as of heat (moxibustion or IR laser) and laser acupuncture with visible light. A key step initiating acupuncture effects is degranulation of mast cells within the connective tissue of acupoints (Zhang et al., 2008, Wang et al., 2014, Wang et al., 2021). The exocytotic process of degranulation results in an increase of cell surface area, which can be followed up during whole-cell recording of electrical membrane capacity (for red laser light as stimulus see Fig. 8.26 (two left columns)). A prerequisite for the degranulation is an increase in intracellular Ca^{2+} (Neher, 1988). A combination of electrophysiology with fluorescence microscopy is an efficient method to investigate the underlying mechanisms (Wang & Schwarz, 2012; Wang et al., 2013). Figure 8.26 (four right columns) illustrates for various physical stimuli the increase in intracellular Ca^{2+}-sensitive fluorescence in mast cells.

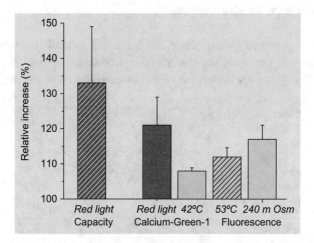

Fig. 8.26 Increase of membrane capacity (whole-mast-cell patch clamp) and calcium-green-1 fluorescence (mast-cell suspension) in response to physical stimuli for at least 5 min laser light of 656.7 nm at 35–48 mW, superfusion of preheated or hypotonic solution. Data represent means + SEM of n = 4–7 experiments and are based on data presented in (Zhang et al., 2012, Wang et al., 2014, Wang et al., 2015)

8.4.1.2 Mast-Cell Degranulation Is Initiated by Ion-Channel Activation

The question arises where the Ca^{2+} is coming from that leads to the mast-cell degranulation. Since we have in the extracellular medium millimolar Ca^{2+} activity (see e.g. Table 2.3), but intracellularly usually only sub-micromolar activity, likely candidates could be ion channels that promote Ca^{2+} entry. The Ca^{2+}-permeable transient-receptor-potential channels of the vanilloid-sensitive TRP family (TRPV) are activated by some physical stimuli. Human mast cells have the TRPV1, TRPV2, and TRPV4 channels expressed (Zhang et al., 2012), which can be activated by visible laser light (Fig. 8.27). TRPV1 and TRPV2 (Caterina et al., 1997, 1999) are known to be activated by higher temperature exceeding 42 and 53 °C, respectively. The sudden increase in intracellular Ca^{2+} activity at 42 and 53 °C shown in Fig. 8.26 can indeed be attributed to these channels.

The degranulation of mast cells involves the release of mediators, including histamine and ATP. In particular, ATP and its metabolic breakdowns are under extensive investigation as mediators for the analgesic effect of acupuncture (Tang et al., 2019; He et al., 2020; Wang et al., 2021). The extracellularly accumulating ATP is rapidly hydrolysed making in addition to ATP the metabolic product adenosine to an important initiating mediator for the final effects of the physical stimuli. These findings stimulate to investigate the role of ATP-activated P2X (Sect. 8.2.2), G-protein-coupled ATP- activated (P2Y), and adenosine-activated (P1) receptors.

8.4.2   Mechanisms in Effected Sites

Pain sensation is governed by excitatory and inhibitory synaptic activity in the brain. Neurotransmission is based on the release of neurotransmitter into the synaptic cleft (compare e.g. Fig. 7.12); on the arrival of an action potential at the presynaptic nerve

Fig. 8.27 Effect of laser irradiation of human mast cells by red (656.7 nm, 48 mW) or green (532 nm, 36 mW) light for at least 5 min on TRPV-mediated current. Data represent means + SEM, $n = 5$–15 based on Gu et al., 2012, Zhang et al., 2012). The value for TRPV4 is taken from unpublished data by Anna Kutschireiter

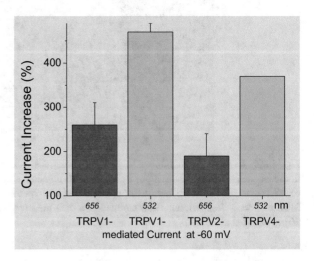

ending neurotransmitter is released in response to Ca^{2+} entry. The transmitter will subsequently bind to and activate transmitter-specific ionotropic and/or G-protein-coupled receptors at the postsynaptic membrane. This will result in depolarisation or hyperpolarisation of the membrane, depending on whether an excitatory or inhibitory transmitter, respectively, is released at the synapse. The dominating excitatory and inhibitory transmitters in the mammalian central nervous system (CNS) are glutamate and gamma-amino-butyric acid (GABA). Both the glutamatergic and the GABAergic system can modulate pain sensation (see e.g.(Goudet et al. 2009)). To terminate synaptic transmission, the neurotransmitter needs to be removed, and this is achieved by high-efficient, Na^{+}–gradient-driven neurotransmitter transporters in the presynaptic neuron and the surrounding glia cells. The dominating transporter for glutamate is the excitatory amino acid carrier 1 (EAAC1), and for GABA the GABA transporter 1 (GAT1, see Sect. 7.1.5 and 8.1.2). The activity of these transporters determines to a large extent the concentration and dwell time of the respective transmitter in the synaptic cleft. Therefore, neurotransmitter transporters play a key role in the regulation of synaptic transmission.

Endogenous morphines, the endorphins, play another important regulatory role for controlling pain sensation, and it was demonstrated that acupuncture results in an increased release of endorphins (Han, 2004). The question, therefore, arises whether activation of opioid receptors can interfere with EAAC1 and GAT1 to promote analgesia.

The activity of the Na^{+}-gradient-driven neurotransmitter transporters, expressed in *Xenopus* oocytes, can be detected by measuring the rate of uptake of radioactive-labelled neurotransmitter or under voltage clamp as current generated by the electrogenic EAAC1 and GAT1. The presence of the δ-opioid receptor (DOR) already reduces the activity of EAAC1 as well as GAT1; this could be demonstrated by co-expression of the respective transporter and DOR (see Fig. 8.28); co-localisation of the transporters and DOR could also be demonstrated by immunofluorescence microscopy and co-immunoprecipitation in *Xenopus* oocytes and also in neurons (Xia et al., 2006, Pu et al., 2012, Schwarz & Gu, 2013) suggesting direct interaction between the different proteins.

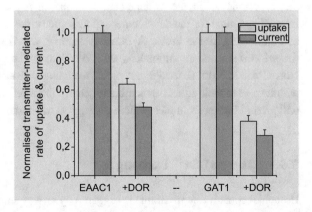

Fig. 8.28 Effects of co-expression of neurotransmitter transporter EAAC1 or GAT1 with DOR in *Xenopus* oocytes. Data represent means + SEM of at least 6 oocytes and are based on results from Xia et al. (2006) and Pu et al. (2012)

Fig. 8.29 Effect of stimulation of EAAC1- and GAT1-mediated current at −60 mV in *Xenopus* oocytes by stimulation of co-expressed DOR (based on data from Xia et al. 2006 and Deng et al., 2009)

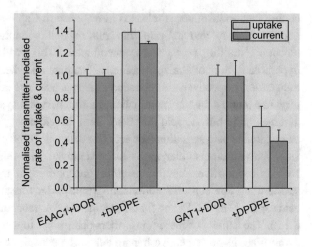

8.4.2.1 Co-Expression of Neurotransmitter Transporters and δ-Opioid Receptor

The DOR can selectively be activated by [D-Pen^2,D-Pen^5]enkephalin (DPDPE, a synthetic opioid peptide). Stimulation of DOR has opposite effects on the two neurotransmitter transporters, while activity of EAAC1 becomes enhanced, GAT1 activity is attenuated (Fig. 8.29). These opposite effects could complement each other in analgesia; elevated activity of EAAC1 will lead to reduced concentration of the excitatory glutamate in the synaptic cleft, and the reduced activity of GAT1 to elevated concentration of the inhibitory GABA. Stimulated glutamatergic activity (see Zhang et al. 2002) and attenuated GABAergic activity (Hu et al., 2003) results in reduced pain sensation (see also Schwarz & Gu, 2013).

8.5   Electrophysiology as a Tool in Pharmacology

As we have illustrated already above for the viral ion channels (see Sect. 8.3), electrophysiology provides with the voltage-clamp technique a useful tool for pharmacology. An interesting source for the development of new medicines turned out to be drug components of traditional Chinese herbs. A famous example was the discovery of artemisinin from *Artemisia annua* as an antimalaria drug by Youyou Tu (2011), which was honoured by Nobel Prize in 2015. Electrophysiological techniques are also useful methods to screen drugs with respect to their effect on membrane protein function. Therefore, in this final section we will present further examples of herbal drugs on membrane transporters and ion channels.

8.5.1   The Na$^+$,Ca^{2+} Exchanger

Bronchial constriction in asthma is triggered by airway smooth muscle (ASM) contraction. Hyperactivity of muscle cells depends on intracellular Ca^{2+} activity (Ca^{2+}$_i$), which is to a large extend controlled by the Na$^+$,Ca^{2+} exchanger NCX1, a secondary active, Na$^+$-gradient-driven transporter for Ca^{2+}, which operates most likely in a 3:1 Na$^+$:Ca^{2+} exchange mode (Hinata & Kimura, 2004). Therefore, it is electrogenic and its activity can be monitored under voltage clamp. In asthma, hyperactivity of NCX1 leads to elevated Ca$^{2+}_i$ operating in reversed mode transporting Ca^{2+} into the cell in exchange to Na$^+$ (Hirota et al., 2007; Janssen, 2009). The reversed mode can be detected as an inward-directed current (comp. Figure 8.30). In *Xenopus* oocytes with expressed NCX1 this current can be extracted as current blocked by Ni^{2+} (see Laudenbach et al., 2017). Drugs that inhibit the exchanger have been discussed for treatment of asthma (see Laudenbach et al., 2017). We will now provide two examples for that.

In TCM, in addition to application of herbal drugs acupuncture has been used to treat asthma. It could be demonstrated in rat animal model that acupuncture leads to elevated levels of the protein cyclophilin A (CyPA) in blood plasma and relief from asthma symptoms (Wang et al., 2009). Therefore, the question arises whether the anti-asthmatic effect of CyPA may be attributed to inhibition of the NCX1. Indeed, incubation of *Xenopus* oocytes with heterologously expressed NCX1 in 4.8 μM CyPA led to strong inhibition of the NCX1-mediated (Fig. 8.30).

In TCM, extracts from roots of liquorice are believed to promote relief from asthma (compare Laudenbach et al., 2017), and glycyrrhizic acid (GA) is a major component. Indeed, similar degree of inhibition of NCX1-mediated current as with CyPA at 4.8 μM occurs at 20 μM with GA (see Fig. 8.30). In contrast to CyPA, the effect develops immediately after application of the drug containing solution. This suggests that the effect of CyPA may be due to a down-regulation of transport activity while GA blocks by direct interaction.

Fig. 8.30 Voltage dependencies of NCX1-mediated current in *Xenopus* oocytes determined as current blocked by 2 mM Ni^{2+}, open circles in the absence of drugs, blue squares in the presence of 4.8 μM CyPA, and green circle of 20 μM GA (based on data by Laudenbach et al., 2017)

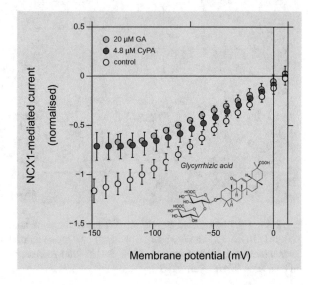

8.5.2   Neurotransmitter Transporters

The dominating neurotransmitter transporters GAT1 (inhibitory) and EAAC1 (excitatory) (see also Sects.7.1.5, 8.1.2, 8.4.2) play a crucial role in the central nervous system and are involved in physiological and pathophysiological function. Therefore, they are important targets for pharmacological treatment of various diseases. For example, drugs have been developed against epilepsy that inhibit the GABA transporter GAT1; this will lead to a prolonged elevated concentration of GABA in the synaptic cleft, and hence, result in increased inhibitory synaptic activity. Tiagabine has successfully been applied as an antiepileptic drug, which selectively inhibits GAT1 (see Eckstein-Ludwig et al., 1999). In Fig. 8.29 we have illustrated that the glutamate transporter EAAC1 can be stimulated by activating DOR; this will lead to reduced dwell time of glutamate in the synaptic cleft, and hence, result in reduced excitatory synaptic activity. Therefore, drugs that stimulate EAAC1 may contribute to pain suppression.

Green tea is a popular drink in east Asia with a long history, and extracts have been used in TCM to treat various diseases. We tested the main and most effective components catechin (−)-epigallocatechin-3-gallate (EGCG) and (−)-epicatechin (EC) with respect to their interference with GAT1 and EAAC1 activity.

While EGCG up to 100 μM has no effect on GAT1-mediated current, 100 μM EC effectively reduced GAT1-mediated current (Fig. 8.31a). Figure 8.31b illustrates that the inhibitory interaction of EC with GAT1 is hardly affected by membrane potential as

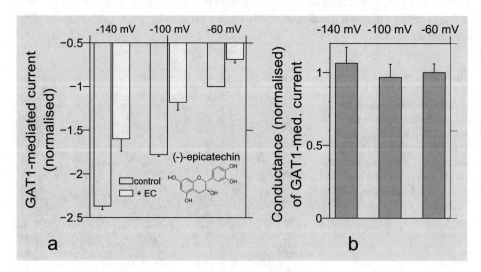

Fig. 8.31 GAT1-mediated current at 3 different clamp potentials (determined as current activated by 100 μM GABA. (**a**) In the absence (control) and presence of 100 μM EC. (**b**) Conductance of the current remaining in the presence of 100 μM EC (data represent means +/− SEM ($n = 5–7$) and based on data from Wang et al. 2016b)

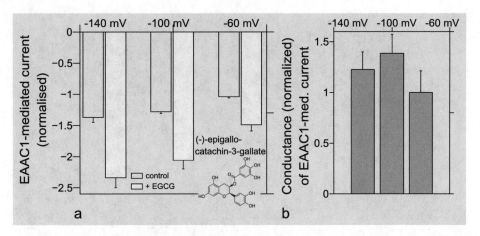

Fig. 8.32 EAAC1-mediated current at 3 different clamp potentials (determined as current activated by 300 µM glutamate. (**a**) in the absence (control) and presence of 100 µM EC. (**b**) Conductance of the current component blocked by 100 µM EGCG (based on data by Wang et al., 2016b)

indicated by the nearly potential-independent conductance of this component in the presence of EC.

In contrast EC has, up to 100 µM, no effect on EAAC1-mediated current, but 100 µM EGCG effectively reduced this current (Fig. 8.32a). Figure 8.32b illustrates that the inhibitory interaction of EGCG with EAAC1 is not significantly affected by membrane potential as indicated by the nearly potential-independent conductance of this component in the presence of EGCG.

8.5.3 Ion Channels

We have already discussed above for viral ion channels (Sect. 8.3) how electrophysiology can serve as an excellent tool to screen for antiviral drug and to investigate their action mechanism. As a final example for electrophysiology in pharmacology, we will pick two ion channels that we have also discussed. The P2X7 receptor, a member of the P2X family (Sect. 8.2.2), plays a role in proinflammatory processes (see e.g. Carroll et al. 2009), and channels of the TRPV family that play crucial role in initiating acupuncture effects by inducing mast-cell degranulation (Sect. 8.4.1).

The P2X7 receptor channel is expressed in inflammatory cells, including mast cells (Jacob et al., 2013), and is involved in mast-cell degranulation (Shen et al. 2019), and hence is involved in transmission of information from the local site to distant areas including central nervous system. The drugs we tested were also used in Chinese herbal therapy. While a crude extract from *Rosmarinus officinalis* stimulates current through P2X7 channels, α-asarone from *Acorus gramineus* as well as a crude extract inhibit the current

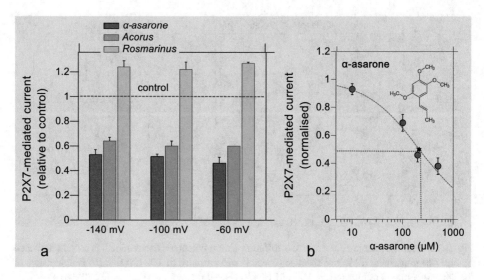

Fig. 8.33 (**a, b**) Effects of herbal drugs on P2X7-mediated current measured in whole-cell configuration of human mast cell line as current activated by 1 mM ATP (data represent means +/± SEM and are based on Spielmann et al., 2008)

(Fig. 8.33a); the effects do not depend on membrane potential. The concentration dependence of the most efficient α-asarone reveals 50% inhibition at about 200 μM (Fig. 8.33b.)

We have described in Sect. 8.4.1 that activation of TRPV1 by physical stimuli contributes to mast-cell degranulation, and hence it may initiate acupuncture effects. Since acupuncture is supplemented in TCM by treatment with herbal extracts, drugs from such herbs may also be used in western medicine to treat, e.g. pain, and TRPV1 could be a target. Extracts from *Evodia ruteacarpa* have been used in TCM to treat different types of pain, and evodiamine and eutaecarpine extracted from fruits of *Evodia* are effective inhibitors of TRPV1-mediated current as demonstrated in patch-clamp experiments on TRPV1-transfected HEK-293 cells (Wang et al. 2016a). TRPV1 is known as receptor for capsaicin responsible for the burning sensation of Chili pepper (see e.g. Darré and Domene (2015)). In *Xenopus* oocytes with expressed TRPV1 the channel characteristics can be investigated after activation by capsaicin. We screened some drugs, known from Chinese herbal medicines, for their effect on TRPV1-mediated current.

"Dragon's blood" obtained from *Dracaena cochinchinensis* is a renowned analgesic in TCM and is significantly effective for treating diseases such as pelvic pain and pain from intercourse (Gupta et al., 2008). One of the main components is loureirin B, which plays a primary role in the pharmacological effect of Dragon's blood (Wei et al., 2013), and TRPV1 was considered to be the target for the analgesic effect. The capsaicin-induced

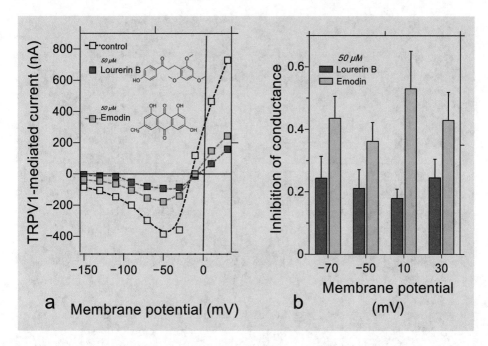

Fig. 8.34 Inhibition of TRPV1-mediated current by 50 μM loureirin B and 50 μM emodin. Current activated by 500 nM capsaicin is considered as TRPV1-mediated current. (**a**) Current–voltage dependencies in the absence and presence of loureirin B or emodin. (**b**) Relative inhibition of conductance by loureirin B and emodin at different membrane potentials (based on unpublished data by Yi Hu, Di Zhang, and Wolfgang Schwarz)

current through TRPV1 channels shows pronounced outward rectification with a reversal potential close to -10 mV (Fig. 8.34a). The reduction of the conductance by 50 μM Loureirin B is not significantly affected by membrane potential (Fig. 8.34b) indicating that the inwardly rectifying voltage dependence is not affected.

Emodin is not only an effective anti-coronaviral drug (see Sect. 8.3.1), but also several analgesic Chinese herbal medicines abound in emodin. Figure 8.34 demonstrates, like loureirin B, voltage-independent inhibition of the conductance though less effective at 50 μM.

In conclusion, we wanted to illustrate that electrophysiology is an effective tool in pharmacology to screen drugs and elucidate underlying mechanisms for their effectiveness in treating diseases by screening for their efficiency in modulating the function of membrane transport proteins. Based on this knowledge, the structure of these drugs can form the basis for development of new medicines.

Take-Home Messages
1. Electrophysiology is a **powerful technique to identify and characterise functionally** (physiologically and pathophysiologically) **significant polypeptide domains and residues**

    e.g.

    Na^+,K^+-ATPase: amino acids involved in formation of access channel or the 2nd gate.

    GAT1: amino acids involved in modulation of activity by glycosylation.

    P2X: Sensitivity and specificity for extracellular ligands and functional consequences of N-glycosylation.
2. Activity of **ion-selective channels** plays an essential role in **virus replication.** Therefore, inhibitors of **viral-channel** activity are potential **antiviral drugs**

    e.g.

    Inhibition of 3-protein of SARS-CoV inhibits virus release from infected cells.

    Highly effective **inhibitors of various viral ion channels** are **kaempferol glycosides**.
3. Electrophysiology is a **powerful technique to investigate** cellular mechanisms underlying **effects of TCM treatment**

    e.g.

    Initiating basis in acupuncture is physical-stress-induced degranulation and release of ATP.

    Regulation of neurotransmitter transporter activity by opioid receptor activation.
4. Electrophysiology is in general a **powerful technique to investigate** the basis of **effects of drug** that can form the basis for development of medicines.

Exercises

1. What are the major tasks of the sodium pump?
2. What are the functional meanings of the α-, ß-, and γ-subunits of the sodium pump?
3. What is the physiological function of the P2X receptors?
4. What role does a neurotransmitter transporter play in physiological and pathophysiological functions?
5. Which ion-channel families do you know?
6. Describe the role of viral ion channels in virus reproduction.
7. How can electrophysiology serve as a tool in development of antiviral drug? Name examples.

References

Bagriantsev, S. N., Gracheva, E. O., & Gallagher, P. G. (2014). Piezo proteins: regulators of mechanosensation and other cellular processes. *Journal of Biological Chemistry, 289,* 31673–31681.

Brake, A. J., Wagenbach, M. J., & Julius, D. (1994). New structural motif for ligand-gated ion channels defined by an ionotropic ATP receptor. *Nature, 371,* 519–523.

Burnstock, G. (1972). Purinergic nerves. *Pharmacological Reviews, 24,* 509–581.

Burnstock, G. (1999). Current status of purinergic signalling in the nervous system. *Progress in Brain Research, 120,* 3–10.

Burnstock, G. (2009). Acupuncture: a novel hypothesis for the involvement of purinergic signalling. *Medical Hypotheses, 73,* 470–472.

Carroll, W. A., Donnelly-Roberts, D., & Jarvis, M. F. (2009). Selective P2X(7) receptor antagonists for chronic inflammation and pain. *Purinergic Signalling, 5,* 63–73.

Caterina, M., Rosen, T., Tominaga, M., et al. (1999). A capsaicin-receptor homologue with a high threshold for noxious heat. *Nature, 398,* 436–441.

Caterina, M., Schumacher, M., Tominaga, M., et al. (1997). The capsaicin receptor: A heat-activated ion channel in the pain pathway. *Nature, 389,* 816–824.

Christensen, A. P., & Corey, D. P. (2007). TRP channels in mechanosensation: direct or indirect activation. *Nature Reviews Neuroscience, 8,* 510–521.

Cordes, F. S., Kukol, A., Forrest, L. R., Arkin, I. T., Sansom, M. S. P., & Fischer, W. B. (2001). The structure of the HIV-1 Vpu ion channel: modelling and simulation studies. *Biochimica et Biophysica Acta (BBA) - Biomembranes, 1512,* 291–298.

Darré, L., & Domene, C. (2015). Binding of capsaicin to the TRPV1 ion channel. *Molecular Pharmaceutics, 7,* 4454–4465.

De Clercq, E. (2006). Antiviral agents active against influenza a viruses. *Nature Reviews Drug Discovery, 5,* 1015–1025.

Delmas, P., & Coste, B. (2013). Mechano-gated ion channels in sensory systems. *Cell, 155,* 278–284.

Deng, H., Yang, Z., Li, Y., et al. (2009). Interactions of Na+,K+-ATPase and co-expressed δ-opioid receptor. *Neuroscience Research, 65,* 222–227.

Ding, S., & Sachs, F. (1999). Single channel properties of P2X2 purinoceptors. *Journal of General Physiology, 113,* 695–720.

Eckstein-Ludwig, U., Fei, J., & Schwarz, W. (1999). Inhibition of uptake, steady-state currents, and transient charge movements generated by the neuronal GABA transporter by various anticonvulsant drugs. *British Journal of Pharmacology, 128,* 92–102.

Fischer, W. B., & Sansom, M. S. P. (2002). Viral ion channels: structure and function. *Biochimica et Biophysica Acta (BBA) - Biomembranes, 1561,* 27–45.

Gereau, R. W., Sluka, K. A., Maixner, W., Savage, S. R., Price, T. J., Murinson, B. B., Sullivan, M. D., & Fillingim, R. B. (2014). A pain research agenda for the 21st century. *The Journal of Pain, 15,* 1203–1214. https://doi.org/10.1016/j.jpain.2014.09.004

Glitsch, H. G. (2001). Electrophysiology of the sodium-potassium-ATPase in cardiac cells. *Physiological Reviews, 81,* 1791–1826.

Goldman, N., Chen, M., Fujita, T., et al. (2010). Adenosine A1 receptors mediate local antinociceptive effects of acupuncture. *Nature Neuroscience, 13,* 883–888.

Goudet, C., Magnaghi, V., Landry, M., Nagy, F., Gereau, R. W., & Pin, J. P. (2009). Metabotropic receptors for glutamate and GABA in pain. *Brain Research Reviews, 60,* 43–56.

Gu, Q., Wang, L., Huang, F., & Schwarz, W. (2012). Stimulation of TRPV1 by green laser light. *Evidence-based Complementary and Alternative Medicine, 2012,* 8. https://doi.org/10.1155/2012/857123

Gupta, D., Bleakley, B., & Gupta, R. K. (2008). Dragon's blood: botany, chemistry and herapeutic uses. *Journal of Ethnopharmacology, 115*, 361–380.

Han, J.-S. (2004). Acupuncture and endorphins. *Neuroscience Letters, 361*, 258–261.

Hao, P., Jiang, F., Cheng, J., Ma, L., Zhang, Y., & Zhao, Y. (2017). Traditional Chinese medicine for cardiovascular disease: evidence and potential mechanisms. *Journal of the American College of Cardiology, 69*, 2952–2966.

He, J. R., Yu, S. G., Tang, Y., & Illes, P. (2020). Purinergic signalling as a basis of acupuncture-induced analgesia. *Purinergic Signal, 16*, 297–304.

Hinata, M., & Kimura, J. (2004). Forefront of Na^+/Ca^{2+} exchanger studies: stoichiometry of cardiac Na^+/Ca^{2+} exchanger; 3:1 or 4:1? *Journal of Pharmacological Sciences, 96*, 15–18.

Hirota, S., Pertens, E., & Janssen, L. J. (2007). The reverse mode of the Na^+/Ca^{2+} exchanger provides a source of Ca^{2+} for store refilling agonist-induced Ca^{2+} mobilization. *American Journal of Physiology-Lung Cellular and Molecular Physiology, 292*, L438–L447.

Hu, J. H., Yang, N., Ma, Y. H., Zhou, X. G., Jiang, J., Duan, S. H., Mei, Z. T., Fei, J., & Guo, L. H. (2003). Hyperalgesic effects of gamma-aminobutyric acid transporter I in mice. *Journal of Neuroscience Research, 73*, 565–572.

Jacob, F., Pérez Novo, C., Bachert, C., & Van Crombruggen, K. (2013). Purinergic signalling in inflammatory cells: P2 receptor expression, functional effects, and modulation of inflammatory responses. *Purinergic Signal, 9*, 285–306.

Janssen, L. J. (2009). Asthma therapy: how far have we come, why did we fail and where should we go next. *European Respiratory Journal, 33*, 11–20.

Kern, D. M., Sorum, B., Mali, S. S., Hoel, C. M., Sridharan, S., Remis, J. P., Toso, D. B., Kotecha, A., Bautista, D. M., & Brohawn, S. G. (2021). Cryo-EM structure of SARS-CoV-2 ORF3a in lipid nanodiscs. *Nature Structural & Molecular Biology, 28*, 573–582.

Krüger, J., & Fischer, W. B. (2009). Assembly of viral membrane proteins. *Journal of Chemical Theory and Computation, 5*, 2503–2513.

Laudenbach, J., Wang, Y., Xing, B. B., Schwarz, S., Xu, Y. F., Gu, Q. B., & Schwarz, W. (2017). Inhibition of the Na^+/Ca^{2+} exchanger NCX1 expressed in Xenopus oocyte by glycyrrhizic acid and Cyclophylin A. *Journal of Biosciences and Medicines, 5*, 128–121.

Liu, Y., Eckstein-Ludwig, U., Fei, J., & Schwarz, W. (1998). Effect of mutation of glycosylation sites on the Na^+ dependence of steady-state and transient current generated by the neuronal GABA transporter. *Biochimica et Biophysica Acta, 1415*, 246–254.

Lu, W., Zheng, B. J., Xu, K., Schwarz, W., Du, L. Y., Wong, C. K. L., Chen, J. D., Duan, S. M., Deubel, V., & Sun, B. (2006). Severe acute respiratory syndrome-associated coronavirus 3a protein forms an ion channel and modulates virus release. *PNAS, 103*, 12540–12545.

Luik, P., Chew, C., Aittoniemi, J., Chang, J., Wentworth, P., Dwek, R. A., Biggin, P. C., Vénien-Bryan, C., & Zitzmann, N. (2009). The 3-dimensional structure of a hepatitis C virus p7 ion channel by electron microscopy. *Proceedings of the National Academy of Sciences, 106*, 12712–12716.

Mandala, V. S., McKay, M. J., Shcherbakov, A. A., et al. (2020). Structure and drug binding of the SARS-CoV-2 envelope protein transmembrane domain in lipid bilayers. *Nature Structural & Molecular Biology, 27*, 1202–1208.

Mould, J. A., Paterson, R. G., Takeda, M., Ohigashi, Y., Venkataraman, P., Lamb, R. A., & Pinto, L. H. (2003). Influenza B virus BM2 protein has Ion Channel activity that conducts protons across membranes. *Developmental Cell, 5*, 175–184.

Neher, E. (1988). The influence of intracellular calcium concentration on degranulation of dialysed mast cells from rat peritoneum. *The Journal of Physiology, 395*, 193–214.

Nicke, A., Bäumert, H. G., Rettinger, J., Eichele, A., Lambrecht, G., Mutschler, E., & Schmalzing, G. (1998). P2X$_1$ and P2X$_3$ receptors form stable trimers: a novel structural motif of ligand-gated ion channels. *The EMBO Journal, 17*, 3016–3028.

Nilius, B., & Honoré, E. (2012). Sensing pressure with ion channels. *Trends in Neurosciences, 35*, 477–486.

Patargias, G., Barke, T., Watts, A., & Fischer, W. (2009). Model generation of viral channel forming 2B protein bundles from polio and coxsackie viruses. *Molecular Membrane Biology, 26*, 309–320.

Pu, L., Xu, N. J., Xia, P., et al. (2012). Inhibition of activity of GABA transporter GAT1 by delta-opioid receptor. *Evidence-Based Complementary and Alternative Medicine: eCAM, 2012*, 12.

Ranade, S. S., Syeda, R., & Patapoutian, A. (2015). Mechanically activated ion channels. *Neuron, 87*, 1162–1179.

Rettinger, J., Aschrafi, A., & Schmalzing, G. (2000). Roles of individual N-glycans for ATP potency and expression of the rat P2X$_1$ receptor. *The Journal of Biological Chemistry, 275*, 33542–33547.

Ruch, T. R., & Machamer, C. E. (2012). The coronavirus E protein: assembly and beyond. *Viruses, 4*, 363–382.

Sakaguchi, T., Tu, Q., Pinto, L. H., & Lamb, R. A. (1997). The active oligomeric state of the minimalistic influenza virus M2 ion channel is a tetramer. *Proceedings of the National Academy of Sciences of the United States of America, 13*, 5000–5005.

Sauter, D., Schwarz, S., Wang, K., Zhang, R. H., Sun, B., & Schwarz, W. (2014). Genistein as antiviral drug against HIV Ion Channel. *Planta Medica, 80*, 682–687.

Schubert, U., Ferrer-Montiel, A. V., Oblatt-Montal, M., Henklein, P., Strebel, K., & Montal, M. (1996). Identification of an ion channel activity of the Vpu transmembrane domain and its involvement in the regulation of virus release from HIV-1-infected cells. *FEBS Letters, 398*, 12–18.

Schwarz, S., Sauter, D., Lu, W., Wang, K., Sun, B., Efferth, T., & Schwarz, W. (2012). Coronaviral ion channels as target for Chinese herbal medicine. *Forum on Immunopathological Diseases and Therapeutics, 3*, 1–13.

Schwarz, S., Sauter, D., Wang, K., Zhang, R. H., Sun, B., Karioti, A., Bilia, A. R., Efferth, T., & Schwarz, W. (2014). Kaempferol derivatives as antiviral drugs against 3a channel protein of coronavirus. *Planta Medica, 80*, 177–182.

Schwarz, S., Wang, K., Yu, W., Sun, B., & Schwarz, W. (2011). Emodin inhibits current through SARS-associated coronavirus 3a protein. *Antiviral Research, 90*, 64–69.

Schwarz, W., & Gu, Q. (2013). Cellular mechanisms in acupuncture points and affected sites. In Y. Xia, G. H. Ding, & G.-C. Wu (Eds.), *Current research in acupuncture* (pp. 37–51). Springer.

Schwarz, W., & Vasilets, L. A. (1996). Structure-function relationships of Na$^+$/K$^+$-pumps expressed in Xenopus oocytes. *Cell Biology International, 20*, 67–72.

Shen, D., Shen, X., Schwarz, W., Grygorczyk, R., & Wang, L. (2019). P2Y$_{13}$ and P2X$_7$ receptors modulate mechanically induced adenosine triphosphate release from mast cells. *Experimental Dermatology, 29*, 499–508.

Shen, D., Zheng, Y. W., Zhang, D., Shen, X. Y., & Wang, L. N. (2021). Acupuncture modulates extracellular ATP levels in peripheral sensory nervous system during analgesia of ankle arthritis in rats. *Purinergic Signalling, 17*(3), 411–424.

Spielmann, A., Gu, Q. B., Ma, C. H., et al. (2008). Inhibition of P2X$_7$ receptor by extracts of Chinese medicine. *Journal of Acupuncture and Tuina Science, 6*, 286–288.

Tang, Y., Yin, H. Y., Liu, J., et al. (2019). P2X receptors and acupuncture analgesia. *Brain Research Bulletin, 151*, 144–152.

Tu, Y. Y. (2011). The discovery of artemisinin (qinghaosu) and gifts from Chinese medicine. *Nature Medicine, 17*, 1217–1220.

Valera, S., Hussy, N., Evans, R. J., Adami, N., North, R. A., Surprenant, A., & Buell, G. (1994). A new class of ligand-gated ion channel defined by P2X receptor for extracellular ATP. *Nature, 371,* 516–519.

Vasilets, L. A., & Schwarz, W. (1993). Structure-function relationships of cation binding in the Na^+/K^+-ATPase. *Biochimica et Biophysica Acta, 1154,* 201–222.

Wang, K., Xie, S., & Sun, B. (2011). Viral proteins function as ion channels. *Biochimica et Biophysica Acta, 1808,* 510–515.

Wang, L., Ding, G., Gu, Q., & Schwarz, W. (2010). Single-channel properties of a stretch-sensitive chloride channel in the human mast cell line HMC-1. *European Biophysics Journal, 39,* 757–767.

Wang, L., Sikora, J., Hu, L., et al. (2013). ATP release from mast cells by physical stimulation: a putative early step in activation of acupuncture points. *Evidence-based Complementary and Alternative Medicine, 2013.* https://doi.org/10.1155/2013/350949

Wang, L.N., Grygorcyk R, Gu, Q.B. and Schwarz, W. (2021) Cellular mechanisms in acupuncture research, In: Advanced acupuncture research 2nd. Ed., editors Ding GH, Shen XY and Wang YQ, Springer

Wang, L. N., Hu, L., Grygorczyk, R., et al. (2015). Modulation of extracellular ATP content of mast cells and DRG neurons by irradiation: studies on u8nderlying mechanism of low-level-laser therapy. *Mediators of Inflammation, 2015,* 9.

Wang, L. N., & Schwarz, W. (2012). Activation of mast cells by acupuncture stimuli. *Forum on Immunopathological Diseases and Therapeutics, 3,* 41–50.

Wang, L. N., Zhang, D., & Schwarz, W. (2014). TRPV channels in mast cells as a target for low-level-laser therapy. *Cell, 3,* 662–673.

Wang, Y., Cui, J. M., Ma, S. L., Liu, Y. Y., Yin, L. M., & Yang, Y. Q. (2009). Proteomics analysis of component in serum with anti-asthma activity derived from rats treated by acupuncture. *Journal of Acupuncture and Tuina Science, 7,* 326–331.

Wang, S., Yamamoto, S., Kogure, Y., Zhang, W., Noguchi, K., & Dai, Y. (2016a). Partial activation and inhibition of TRPV1 channels by evodiamine and rutaecarpine, two major components of the fruits of *Evodia rutaecarpa. Journal of Natural Products, 79,* 1225–1230.

Wang, Y. X., Engelmann, T., Xu, Y. F., & Schwarz, W. (2016b). Catechins from green tea modulate neurotransmitter transporter activity in Xenopus oocytes. *Cogent Biology, 2,* 1261577.

Wei, L. S., Chen, S., Huang, X. J., Yao, J., & Liu, X. M. (2013). Material basis for inhibition of dragon's blood on capsaicin-induced TRPV1 receptor currents in rat dorsal root ganglion neurons. *European Journal of Pharmacology, 702,* 275–284.

Wu, C. H., Vasilets, L. A., Takeda, K., Kawamura, M., & Schwarz, W. (2003). The N-terminus of the Na^+,K^+-ATPase α-subunit acts like an inactivation "ball". *Biochimica et Biophysica Acta, 1609,* 55–62.

Xia, P., Pei, G., & Schwarz, W. (2006). Regulation of the glutamate transporter EAAC1 by expression and activation of δ-opioid receptor. *The European Journal of Neuroscience, 24,* 87–93.

Zhang, D., Ding, G., Shen, X., et al. (2008). Role of mast cells in acupuncture effects: a pilot study. *Explore, 4,* 170–177.

Zhang, D., Spielmann, A., Wang, L., et al. (2012). Mast-cell degranulation induced by physical stimuli involves the activation of transient-receptor-potential channel TRPV2. *Physiological Research, 61,* 113–124.

Zhang, Y.-Q., Ji, G.-C., Wu, G.-C., & Zhao, Z.-Q. (2002). Excitatory amino acid receptor antagonists and electroacupuncture synergetically inhibit carrageenan-induced behavioral hyperalgesia and spinal fos expression in rats. *Pain, 99,* 525–535.

Appendix

9

Contents

Abstract

This book is supplemented by appendices (Chapter 9) describing influence of electrical and magnetic fields on physiological function, and a manual for a Laboratory Course in electrophysiology using the two-electrode voltage-clamp technique.

Keywords

Effect of electrical fields · Effect of magnetic fields · Electro smog · Laboratory course

9.1 Influence of External Electrical and Magnetic Fields on Physiological Function

We have learned that in a living organism electric currents are flowing across and along cell membranes. This became particularly apparent for an action potential travelling along a nerve fibre (see e.g. Fig. 6.20). Another example of current flowing along the cell surface, which can be detected extracellularly, is the *Xenopus* oocyte using an interesting electrophysiological method: the **vibrating electrode** mounted to a piezo-crystal-driven element (Jaffe & Nuccitelli, 1974; Robinson, 1979). Depending on the position of the electrode vibrating perpendicular and close to the surface of the cell, current signals of different amplitude can be detected (compare Fig. 9.1). At the equator the signal nearly disappears suggesting current flow between the poles. In this respect compare also Sect. 3.3.4.

Changes in current flow will result in electromagnetic signals. We have pointed out in Sect. 3.1 that electrical signals can be recorded with electrodes firmly connected to the body surface, and that analysis of these signals can provide us details about electrophysiological function of an organ in the body (see Table 3.1). As an example, we discussed the electrocardiogram (Sect. 3.2). It should be pointed out that also magnetic signals can be detected outside the body, though highly sensitive devices in the form of the SQUID (Superconductance Quantum Interference Device) are necessary.

We also have briefly mentioned in Sect. 3.1 that physiological function can electrically be modulated via electrodes (see Table 3.1), but again, they have firmly to be attached to the body surface as in electrotherapy. We will come back to this topic in Sect. 9.1.2.

9.1.1 Magnetostatic Fields

Though our focus was "electro"physiology, we, at least briefly, want to address the question to which extend magnetic fields can interfere with cellular processes. First let us consider some flux densities of magnetic fields that might be of physiological relevance (see Table 9.1).

The question had been raised to which extent animals can sense external magnetic fields. It could be demonstrated that certain animals (including, e.g. specialised birds, fishes, or turtles) indeed can sense components of the magnetic field of the earth in the µT range

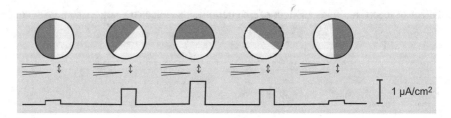

Fig. 9.1 Detection of extracellular current (lower row) perpendicular to the surface of an oocyte in different orientation (upper row) by a vibrating electrode (middle row) (see also Robinson, 1979)

Table 9.1 Magnetic flux densities

Origin	Flux density	
Brain activities	pT - fT	As a consequence of electrical phenomena, detectable with SQUID
Magnetic field at surface of the earth	$\approx 40\ \mu T$	Detectable with compass, by magnetobacteria, and possibly certain animals
Conventional magnet	mT	Household magnet
Magnet for imaging	T	Only in this range biological effect may become of relevance

though underlying mechanisms still need to be elucidated. Up to now there are no indications that biochemical or physiological mechanisms can directly be affected by flux densities in the μT range. Radical pair mechanisms occur only in the mT range, but lifetime of radical pairs can be modulated by magnetic fields in the μT range (see Maeda et al., 2008), which may serve as a refined mechanism in biological magnetic field sensation. The existence of biological magnetits could be demonstrated; but up to now magnetits of functional significance have been found only in magnetobacteria (see e.g. Schüler, 2008).

For "medical treatment" magnets of 100 mT are occasionally advertised, but even flux densities of more than 1 T are not able to affect, e.g. ion movements. This is taken as a strong argument that magnetic resonance tomography is without harm for the human body (see ICNIRP, 1998, 2010, 2014). To modulate fluid movements, the flux densities have to be raised into the range of 10 T.

In conclusion, it seems to be established that highly specialised biological structures exist that allow specialised cells or animals to detect changes in magnetic fields as low as μT. For normal structures, on the other hand, flux densities in the T range are necessary for interaction; this is also manifested by the guidelines of the International Commission on Non-ionizing Radiation Protection (see ICNIRP Guidelines, 1998, 2010, 2014).

9.1.2   Electrostatic Fields

The highly sophisticated electrophysiological techniques, in particular the voltage-clamp technique, allow investigations of the function of membrane proteins that govern cell function. On the other hand, we have also learned that vice versa application of electrical fields to an organ can modulate the function of membrane proteins, as used, e.g. in electroshock therapy or for artificial heart pacemaker (Table 3.1). Even in ancient times, when Roman physicians treated pain by the use of the discharges of the electric organ of *Torpedo*, electrical fields were applied via the body surface (see Sect. 1.2). We like to emphasise again that tight contact of the electrodes with the tissue are essential.

Table 9.2 Static electrical fields

Origin	Intensity
Earth surface field	100 V/m
During thunderstorm	20 kV/m
During flash	1 MV/M
Membrane potential (50 mV)	10 MV/m
Extracellular fields (in 20–30 μM distance)	10 V/m
Stimulation of action potential (0.1 A/m^2)	≈ 0.5 V/m
Wound potential	100 V/m range
Galvanotaxis	≈100 V/m

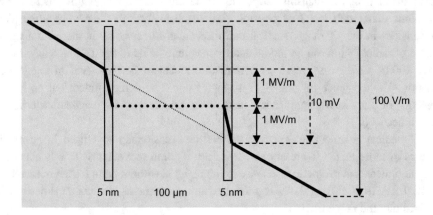

Fig. 9.2 Potential drops in an electrostatic field of 100 V/m at a cell of 100 μm cell in diameter assuming a membrane thickness of 5 nm

Nowadays, effects of electrical fields originating from our environment are enthusiastically discussed in the public as a source for modulating body function, sometimes termed "electrosmog". Though medical applications were successfully introduced, and we cannot think about modern medicine without those inventions, also unrealistic ideas have grown up. Therefore, we should make at least a few clarifying comments on that (Mäntele, 2021).

Table 9.2 lists some typical intensity values of electrical fields from different origins. The earth can be considered as a ball-shaped capacitor that is negatively charged and has a surface field strength E_{earth} of about 100 V/m. During thunderstorm values of 20 kV/m can be reached, during flashes even 1 MV/m.

If E_{earth} of 100 V/m would directly act on the tissue of our body (compare Fig. 9.2), an electrical potential drop across a cell of 100 μm in diameter would be 10 mV; across the cell membranes of 5-nm thickness this field would induce a drop of 5 mV. Accordingly, the field at the membrane would be 1 MV/m.

The potential difference across the membrane of an animal cell amounts to several tens of mV. Assuming a membrane resting potential of an excitable cell of −50 mV and a

Table 9.3 Typical values of specific conductances (compare also Table 2.2)

	Specific conductance(S/m)
Air	$\approx 5 \ 10^{-15}$ S/m
Lipid bilayer	10^{-13} S/m
Tissue	0.3 S/m
Extracellular solution	1 S/m
Sea water	1S/m
Cell membrane	10^3–10^{-2} S/m^2

membrane thickness of 5 nm, the electrical field will amount to $E_{membrane} = 10$ MV/m (Table 9.2).

Field changes in the membrane of physiological relevance, therefore, should be in the range of MV/m. From this point of view electrostatic fields of 100 V/m acting on a cell could be of physiological relevance. To initiate an action potential, current densities of 0.1 A/m^2 will be needed, and during an action potential the current density rises to about 1 A/m^2. Assuming an average specific tissue conductance of about 0.5 S/m (compare Table 9.3) 1 A/m^2 would correspond to 2 V/m.

On the other hand, our body can electrically be represented by an electrolyte container (compare also Sect. 3.2); therefore, the static electrical field will hardly enter the body. Exposing a spherical body with a specific conductance g to an external electric field E_{out} will produce a transient internal field E_{in}:

$$E_{in} = \frac{3\varepsilon_{out}}{2\varepsilon_{out} + \varepsilon_{in}} E_{out} \cdot e^{-t/\tau}$$

with external and internal dielectric constants ε_{out} and ε_{in}, respectively, and time constant $\tau = \varepsilon_{in}\varepsilon_0/g$. For typical parameters ($\varepsilon_{out} = 1$ (air), $\varepsilon_{in} = 100$ (tissue), $g = 0.3$ S/m) we obtain for τ a value of about $2 \ 10^{-9}$ s. According to this fast time constant only electromagnetic field at high frequency can enter body tissue significantly (see below),

Fields in the range of 10 kV/m (as during thunderstorms) may be sensed at the skin surface, but definitively not inside the conducting "electrolyte container". We are facing a different situation in a conductive environment like sea water instead of air (Table 9.3). Certain sea water animals have developed receptor systems (ampullae of Lorenzini (e.g. Fields, 2007)) that allow sensation of fields as low as μV/m (Murray, 1962).

9.1.3 Electromagnetic Fields

Electromagnetic field in particular in the high frequency range (see Fig. 9.3) of course will enter the body, and therefore, the question may be raised to which extend energy can be absorbed by the tissue and whether the electromagnetic fields surrounding us can be harmful to human health and considered as electrosmog.

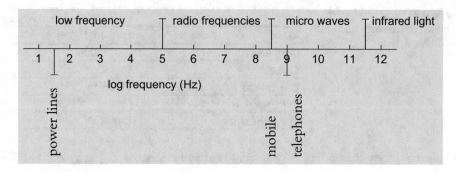

Fig. 9.3 Spectrum of frequencies of technically used electromagnetic wave

9.1.3.1 Low-Frequency Electric Fields (50 Hz)

Due to very high conductivity of tissue compared to air (see Table 9.3) and the fast time constant (see above), low-frequency fields (like the 50 Hz of power supplies) will be in tissue still considerably lower compared to the external field.

Under a high-voltage power line electric fields of 10 kV/m can exist in vertical direction, which may result in a human grounded through his feet in current densities at the surface of 0.6–20 mA/m^2, increasing from head to feet (see Vistnes, 2001). Because initiation of e.g. an action potential needs current densities of at least 100 mA/m^2 (see above), sensations even at the surface may be excluded.

9.1.3.2 High Frequency Electric Fields (kHz - GHz)

A different situation we are facing in the high frequency range (compare above time constant τ), and effects on cellular events may not be excluded. Since different tissues will contribute significantly different to absorbance of electromagnetic energy, the effects on the cellular level of electromagnetic fields are difficult to be estimated. As a first approach, models and methods of analysing simulations have been developed. Upper limits for effects of the human body have been elaborated that are far beyond the values detectable in our environment (ICNIRP, 1998, 2010, 2014).

9.1.3.3 Conclusion

Electrical as well as magnetic fields play important roles in the investigation of cellular, organ, and body function as well as in medical diagnosis and therapy. On the other hand, electric and magnetic fields from environmental source can hardly affect physiological or biochemical function. As we have discussed above, effects of the magnetic field component can be neglected, only fields in the T range may be come of significance. If electrodes are not in direct contact with the tissue, only high frequency electromagnetic field will enter the body, and strong regulations have been elaborated to avoid harm to the body.

9.2 A Laboratory Course: Two-Electrode Voltage Clamp (TEVC)

This appendix presents a description of a manual for a laboratory course on electrophysiology applying TEVC to *Xenopus* oocytes (Fig.9.4) that is based on a manual prepared by Bierwirtz and Schwarz (2014) for courses held at Goethe-University Frankfurt and Fudan-University Shanghai.

Here we will review several topics that were dealt with in this book. The aim of such a course is to exercise the various steps needed to perform an experiment. Here we will use as an example the *Turbo TEC* voltage-clamp system and *CellWorks* software of NPI electronic (Tamm, Germany, for details see **www.npielectronic.com**). Another aim is to learn the basics of a typical protocol of voltage-clamp experiment, which includes separation of different current components and validation of the procedure. The validation of the protocol

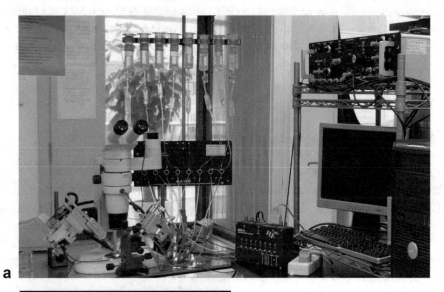

a

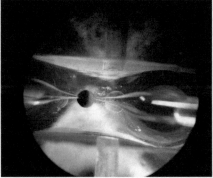

b

Fig. 9.4 Set-up for Two-Electrode Voltage clamp (**a**) using the *Turbo TEC* system from NPI electronic (www.npielectronic.com)) of an oocyte in the voltage-clamp perfusion chamber with two microelectrodes impaled (**b**)

to extract and analyse a specific conductance pathway is a very important procedure. The work of Hodgkin and Huxley (1952) (see also Sect. 6.1) was an excellent example.

9.2.1 Motivation

The function of a cell is governed to a large extent by transport across the cell membrane that is mediated by specific membrane proteins. To learn about functional characteristics and regulation of these membrane proteins, electrophysiological techniques have turned out to be a powerful method. Particularly the combination of electrophysiology with molecular biology and biochemistry allows obtaining fundamental information on structure, function, and regulation of charge-translocating proteins (see Chap. 8). This can be achieved by expression of genetically modified proteins in *Xenopus* oocytes and functional characterization by electrophysiological methods. For simplicity, this laboratory course will be restricted to non-injected oocytes, which is sufficient for the above listed aims.

9.2.2 Background

9.2.2.1 Electrical Characteristics of Biological Membranes

The Membrane Potential
Most electrical phenomena occurring at a cell membrane are based on the asymmetrical ion distributions between cytoplasm and extracellular space (Table 2.3) and on ion-selective membrane conductances (compare Sect. 5.1.1). The electrochemical gradient across the cell membrane leads to an electrical potential difference, the so-called membrane potential that can be detected by electrodes in the intra- and extracellular space. In order to describe this potential, different approaches can be made:

1. If we assume that a membrane is **permeable for all ions except for one species**, the membrane potential can be described by the **Donnan equation** (see Sect. 2.3.1). For instance, macromolecules like anionic proteins or nucleic acids are not able to cross the membrane. For an animal cell, the calculated Donnan potential does not match the actual membrane potential (Table 2.4).
2. Therefore, another approach needs to be chosen. If we assume that a membrane is **impermeable for all ions except for one species**, the membrane potential can be described by the **Nernst equation** (see Sect. 2.3.2):

$$\Delta E = \frac{\mathrm{RT}}{\mathrm{zF}} \ln \left(\frac{a_{\mathrm{out}}}{a_{\mathrm{in}}} \right) \tag{9.1}$$

with R is the universal gas constant, T the absolute temperature, z the valence of the ion, F the Faraday constant, and a the ion activity inside and outside the cell, respectively.

3. For a real cell neither the Donnan nor the Nernst equation can describe the resting membrane potential (Table 2.4) because the membrane has specific permeabilities for the different ion species, and these permeabilities depend on particular environmental conditions. The **Goldman–Hodgkin–Katz equation** is often used to describe the dependence of membrane potential on different ion permeabilities; if Na^+, K^+, and Cl^- were the permeable ions (see Sect. 2.4), the Goldman–Hodgkin–Katz potential is given by.

$$E_{GHK} = \frac{RT}{F} \ln \left(\frac{P_{Na}[Na]_{out} + P_K[K]_{out} + P_{Cl}[Cl]_{in}}{P_{Na}[Na]_{in} + P_K[K]_{in} + P_{Cl}[Cl]_{out}} \right). \tag{9.2}$$

However, to derive Eq. (9.2) three assumptions have to be made:

- independent ion movement (i.e. free diffusion),
- constant diffusion coefficient D_C (homogenous membrane phase),
- constant electrical field within the membrane (linearly changing potential).

For realistic biological conditions these three assumptions are all highly questionable.

The Membrane as an Electrical Unit

Electrical current within a biological system is mediated by ion movements. From an electronic point of view, a biological membrane can be considered as a parallel circuit of a resistance and a capacitance (Fig. 2.1). Opening and closing of ion-permeable channels and activities of electrogenic carriers govern the resistance of the membrane; therefore, current–voltage (IV) characteristics can tell us a lot about the function of the respective membrane proteins. Interestingly, the specific capacitance of a cell membrane hardly changes and is independent of the cell type (see Sect. 2.1). The specific capacitance of a lipid bilayer is close to $0.8 \, \mu F/cm^2$, and a value of $1 \, \mu F/cm^2$ can be used to calculate the surface area of a cell by electrical determinations of the cell capacitance. The capacitance can be calculated from the transient signal of charging or discharging the membrane capacitor.

Applying a small rectangular voltage-clamp pulse to the membrane the current response is composed of a transient, exponential capacitive signal (I_{cap}), and a constant ohmic current (I_{res}) carried by ions crossing the membrane (see Fig. 9.5):

$$I(t) = I_{cap} + I_{ss} = C \frac{dU}{dt} + I_{ss}. \tag{9.3a}$$

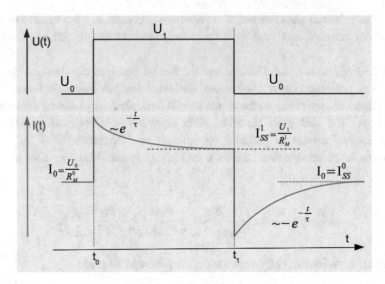

Fig. 9.5 Time course of current in response to a voltage-clamp pulse. The signal is a superposition of a transient (capacitive) and a stationary (resistive) current component, which is ohmic for a small voltage step (I_{SS}) (see Eq. (9.3a))

The membrane capacitance can then be calculated according to:

$$\int_{t_0}^{t_1} (I(t) - I_{SS})dt = C_M \int_{U_0}^{U_1} dU = C_M(U_1 - U_0). \tag{9.3b}$$

Theoretical Background of Voltage Clamp

The most powerful electrophysiological method for basic research is the voltage-clamp technique. The method allows, at a given membrane potential, measurement and analysis of currents across the cell membrane, which are mediated by specialised channels and carriers. The voltage-clamp technique was the basis for the two milestones in modern electrophysiology: the Hodgkin and Huxley (1952) description of excitability (Sect. 6.1) and the demonstration of single-channel events by the patch-clamp technique (Neher and Sakmann (1976), Section 4.4).

The Principle of Voltage Clamp (See Sect. 3.4.5)

Ideal voltage clamp (Fig. 3.19), with electrode resistance ($R_E = 0$), consists of a voltage source providing the clamp potential V_C, the model membrane, a switch and an ampere meter for measuring membrane current I_M. This circuit is "ideal" because wires, ampere meter, and battery are assumed to have negligible intrinsic resistances. Therefore, when the switch is closed, the potential across the model membrane V_M reaches V_C as soon as the capacitor is charged.

The main difference between ideal and real voltage clamp is that the connection between electronic circuit and cell cannot be treated with negligible resistances. In particular, the electrode resistance R_{CE} cannot be neglected, and the two resistors, R_{CE} and R_M, are in series and act as voltage divider. The potential drop across the membrane, therefore, is only:

$$V_M = \frac{R_M}{R_M + R_{CE}} V_C. \tag{9.4}$$

If R_{CE} cannot be neglected compared to R_M, a second electrode is needed to determine the actual membrane potential (see Fig. 3.20). In the laboratory course glass microelectrodes are used to penetrate the cell membrane to access the cell interior. These electrodes have tip resistances in the range of 1–5 MΩ, which is similar to the input resistance of large cells such as the *Xenopus* oocytes.

Two-Electrode Voltage Clamp

For large cells with low input resistances ($R_M \leq R_E$) we use the so-called Two-Electrode Voltage Clamp (TEVC). Since membrane resistance R_M changes in response to various stimuli during an experiment, it is necessary to continuously compare the membrane potential V_M, measured via the potential electrode P_E, with the command potential V_C. This is achieved by electronic devices that allow exact and rapid communication between command and measured membrane potential. The central part of such an electronic set-up is the operational amplifier (*op-amp*) (Fig. 3.22).

The main characteristic of an op-amp (Fig. 3.22a) is its ability to amplify the difference ($e_+ - e_-$) between its two inputs by a gain factor A.

$$e_0 = A(e_+ - e_-). \tag{9.5}$$

This kind of op-amp forms as negative feedback amplifier the central part of the voltage-clamp system (see Fig. 3.23). The positive input is connected to the command potential V_C, the negative input to the signal provided by the potential electrode. These two input signals define the potential at the output, and hence allow the cell to be clamped fast and accurately to the command potential. The current flow from the feedback amplifier is identical to the membrane current and can be measured either at the output of the op-amp or at the grounded bath electrode.

Another essential op-amp of the voltage-clamp system is the voltage follower (Fig. 3.22b) where the negative input is directly connected to the output (i.e. $e_0 = e_-$). At the typically high gain factor (A $\approx 10^4$–10^6), according to Eq. (9.5), the output signal will follow the signal at the positive input ($e_0 \approx e_+$). The voltage follower is used to uncouple the sensitive signal of the voltage electrode from the following recording devices, and to serve as a high resistance input in order to minimize current flow through the voltage electrode.

Very often two bath electrodes are used, one current-passing grounded electrode and one bath electrode serving as a reference electrode for the intracellular voltage electrode (virtual ground). The use of the virtual-ground electrode has the advantage that this bath electrode will not polarize due to current flow.

9.2.3 Questions to Be Answered for the Course

In preparing for the laboratory course the student should be familiar with the following questions:

(a) What are the approximated activities of intra- to extracellular ions for a vertebrate cell?
(b) What is the specific membrane capacitance (capacitance per unit area), and why is it an important biophysical parameter? What is the expected total capacitance of a *Xenopus* oocyte with a diameter of 1 mm (if the oocyte were assumed to have the shape of a perfect sphere)?
(c) Calculate the Nernst potential for a typical K^+ distribution at a biological membrane. How can the Nernst equation be used to determine the intracellular K^+ concentration?
(d) Write down the Goldman–Hodgkin–Katz equation. What are the assumptions, the GHK equation is based on? Give examples that are not compatible with GHK equation.
(e) Gather some information about the Na^+/K^+-ATPase (sometimes also referred to as "sodium pump"), and explain why the ion transport mediated by this ion pump is electrogenic.
(f) Describe the main features of the voltage-clamp technique. How does it work (circuit diagram)? What is the function of an operational amplifier and a voltage follower?
(g) In the laboratory course you have to prepare microelectrodes. How is the electrical connection between solution and the electronics achieved? Give the reaction formula.

9.2.4 Set-up and Basic Instructions

9.2.4.1 Experimental Set-up *(See* Fig. 9.6*)*
Defolliculated oocytes of the clawed toad *Xenopus laevis* are positioned in the cell chamber (1) that is mounted under a binocular and connected to a stopcock or set of valves for changing the solutions perfusing the chamber. The cell is impaled with two microelectrodes (see also Fig. 9.4).

A clamp potential V_C is applied to the membrane by the voltage-clamp amplifier Turbo TEC (NPI electronics, (www.npielectronic.com)) (2). The amplifier is under control of a personal computer (3) on which the program CellWorks is running. For control of the quality of the voltage clamp, the time courses of membrane current I_M and membrane potential V_M are monitored on an oscilloscope (4). In addition, holding potential and holding current are continuously recorded by a pen recorder (5).

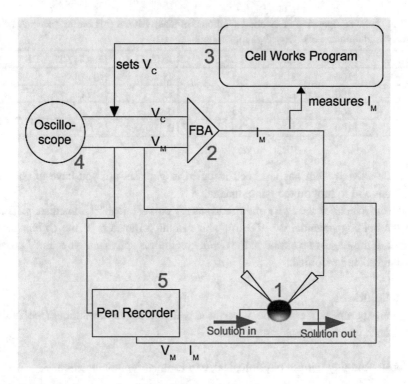

Fig. 9.6 Schematic diagram of experimental set-up (see text)

9.2.4.2 Preparation of Microelectrodes

First, glass micropipettes are pulled from filament-containing capillaries of borosilicate glass using a vertical puller. The pipettes are filled with 3M KCl using a syringe with a fine needle. It is crucial to eliminate all air bubbles from the tip of the microelectrode. The electrodes are inserted into an electrode holder mounted to a micromanipulator. A silver wire provided by the electrode holder and covered with AgCl serves as connection to the electronics. A resistance in the range of 1 MΩ can be check by an "ohmmeter" built-in to the Turbo TEC amplifier.

9.2.4.3 Instructions for the Use of CellWorks Program for the Turbo TEC

1. After starting CellWorks choose the appropriate user (e.g. DEF).
2. Open the following modules.
(a) *"Execution"* (executing valve setting and the pulse protocol for determination of IV curves)
(b) *"Online A"* (displaying time course of the VC pulses, steady-state potential and current, and IV relationship).
3. If a valve system is installed: Run from *Execution* the *Manual* option. Open the valve for the desired solution and the pump will start running.

Table 9.4 Ionic composition of external solutions (in mM). MOPS (pH buffer, pH = 7.2), TMA: tetramethylammonium, TEA: tetraethylammonium

Solution #	NaCl	CaCl$_2$	MOPS	TMA-Cl	BaCl$_2$ / TEA-Cl	KCl
1	100	1	5	25	0 / 0	10
2	100	1	5	25	0 / 0	0
3	100	1	5	0	5 / 20	10
4	100	1	5	0	5 / 20	0

    If the valve system has not been installed to your set-up, you have to change the solutions and to turn on the pump manually.
4. Run from *Execution* the *Pulse* option and start *IV curve*, but abort immediately. This will set the holding potential of −60 mV and default parameters in the *Online* monitor. Check, if the *Export on* option in the *Online* monitor is activated. Now an IV curve can be initiated and recorded.

9.2.4.4 Solutions
During the experiment, the chamber is perfused with four different solutions (see Table 9.4) that influence various current components.

Solution 1 Roughly mimics physiological, extracellular ion composition.

Solution 2 KCl is missing, and hence, current mediated by the (endogenous) Na/K pump, I_P, and inward current through K$^+$ channels, I_K, should be blocked.

Solution 3 TMA-Cl is replaced by BaCl$_2$ and TEA-Cl. Ba^{2+} and TEA$^+$ are applied as specific inhibitors of currents through K$^+$ channels.

Solution 4 the KCl of solution 3 is missing, and hence, current mediated by the Na/K pump should again be blocked.

9.2.5   Experiments and Data Analysis

9.2.5.1 IV Characteristics

Procedure
Since we are interested in detecting current mediated by Na$^+$/K$^+$ pump, we choose conditions that stimulate pump activity. That is why we use extracellular solution with 10 mM K$^+$. The intracellular Na$^+$ concentration was increased by incubating the cells in Na$^+$−loading solution for about 30 min in order to maximize the Na$^+$-pump-mediated current. The loading solution did not contain Ca^{2+}, which destabilizes the negatively

charged membrane surface (see Sect. 6.4.1) and allows equilibration of intracellular and extracellular small cations. Thereafter, the oocytes are placed into a post-loading solution for membrane recovery for at least 15 min before the experiment is started.

After the two microelectrodes have been immersed into the bath solution, the perfusion system should be adjusted to about 1 drop/s for each solution. In the VC-off mode of the Turbo TEC the voltage offsets of potential and current electrode should be adjusted to zero before they are finally inserted into the oocyte. For detecting impalement, it is helpful to turn on the built-in audio monitor of the Turbo TEC, which converts the value of the potential at the electrode tip into an acoustic signal. Once the tip penetrates the cell membrane, the resting potential is indicated by a change in audio frequency.

After both microelectrodes have been inserted properly, the amplifier can be set to *VC* mode. Before starting measurements of IV dependencies (by starting *IV curve* in the CellWorks *Execution* window) the holding current (that is monitored on a pen recorder) should have stabilised. Now the cell can be superfused successively with the different solutions in the following sequence:

1	→	*2*	→	*1*	→	*3*	→	*1*	→	*3*	→	*4*	→	*3*
I_1		I_2		I_3		I_4		I_5		I_6		I_7		I_8

After having changed to another solution, wait for about 1–2 min to allow complete exchange before recording the respective IV dependencies (I_n).

For each measurement the program applies from the holding potential of −60 mV rectangular voltage pulses from −150 mV to +30 mV in 10 mV increments (of 200 ms duration) to the membrane and records the steady-state values of membrane potential and current close to the end of the test pulses, i.e. after the capacitive current component has vanished (as shown in the example presented in Fig. 9.7). Data are averaged over a time period of 20 ms to correct for possible 50-Hz noise.

After having recorded all 8 IV dependencies (IV1 to IV8) in the respective solutions, the experiment should be repeated with several additional oocytes. For final data evaluation at least five successfully performed experiments should be available.

Tasks

For data analysis, the use of software is recommended that provides features such as data handling (e.g. copy/paste), calculation of mean values and errors, plotting, fitting, and hypothesis testing. In the laboratory course, the application of ORIGIN software will be demonstrated.

The date files can be found in the *Export* folder in the CellWorks directory. Every time an IV dependency was recorded, a new ASCII file was created by CellWorks. Each file consists of six columns; only the first (potential in mV) and the second (current in nA) column are needed.

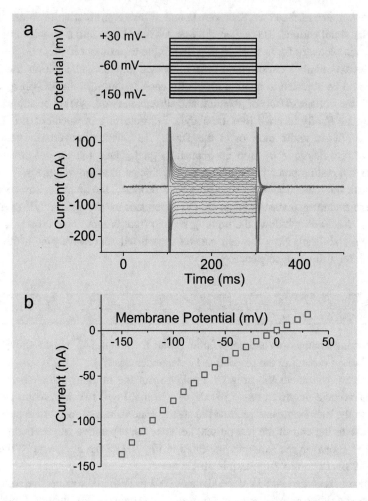

Fig. 9.7 (**a**) Current traces in response to the eliciting potential steps. The steady-state values are obtained from data averaged during the last 20 ms at the end of the pulse. (**b**) Corresponding steady-state IV dependency

For each experiment the current components $I_{K\text{-}sens}$, I_K, and I_{Pump} are to be determined by taking differences of the IV dependencies (see below). Thereafter, the mean values and SEMs (standard error of the mean) should be calculated. The calculated IV characteristics of the single experiments are to be plotted into one graph, and the mean current with SEM as error bar into a separate one.

(a) The total K$^+$-sensitive current $I_{K\text{-}sens}$ is determined from the difference between the currents measured in solution 1 and 2. Sometimes the current during the experiment tends to drift linearly in time. Therefore, a drift correction is performed by taking the arithmetic mean of I_1 and I_3 (both were measured in solution 1):

$$I_{K-\text{sens}} = \frac{I_1 + I_3}{2} - I_2.$$

(b) The K^+ current through the K^+ channels is determined from the difference between the current measured in solution 1 and 3:

$$I_K = \frac{I_3 + I_5}{2} - I_4.$$

(c) The current that was mediated by the Na^+/K^+-ATPase is determined from the difference between the currents measured in solution 3 and 4:

$$I_{\text{Pump}} = \frac{I_6 + I_8}{2} - I_7.$$

(d) To demonstrate that the K^+ channels and the sodium pump are the major sources for K^+-sensitive currents you may compare the sum $I_K + I_{\text{Pump}}$ with $I_{K\text{-sens}}$. Select at least three potentials (for instance -120 mV, -60 mV, and 0 mV) and perform a paired-sample t-test at these potentials (give the t values calculated by the data analysis program in your protocol). Discuss the result.

9.2.5.2 Hypothesis Testing - the Paired-Sample T-Test

For the evaluation of data we determine electrogenic current mediated by the Na^+,K^+-ATPase (I_{pump}), the current through K^+-selective channels (I_K) and total K^+-sensitive ($I_{K\text{-sens}}$) current. These currents were calculated as difference of membrane currents without and with inhibition of the respective current component. The hypothesis is to be tested whether:

$$I_{K-\text{sens}} = I_K + I_{\text{pump}}.$$

In order to distinguish significant differences of currents from statistical deviations a statistical criterion for being significant has to be used.

Fig. 9.8 (a, b) Probability
density function f(t, df) for
df = 3, df = 9, df = 10 and
Gaussian normal distribution.
The blue / red areas mark 5%
areas for the one-tailed t-test
with |t| > 1.83 denoting that the
sample mean is significantly less
/ greater than the test mean. The
striped areas mark the 2.5%
areas for the two-tailed t-test
with |t| > 2.26 denoting that the
sample mean is significantly
different from the test mean

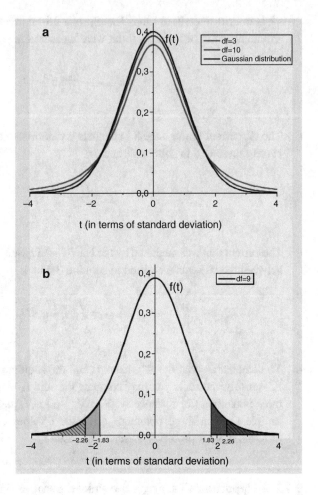

A very detailed introduction into statistical methods in general can be found on the
internet webpage *"Concepts & Applications of Inferential Statistics"* (Lowry, 1999,
vassarstats.net/textbook/).

The t-test is to be applied for testing the above hypothesis. Basis for hypothesis testing is
that the data form a sup-population of an unknown source population that can be supposed
to have normal distribution. The sup-population of the source must have the same mean
value as the source; however, the standard deviation of the source is unknown.

When taking random samples of size N from the source, the mean variance of the
samples is given by the variance of the source population multiplied by $1/N$. The samples
do not follow normal distribution, but rather t-distribution. This distribution is also bell-
shaped, but depending on its degrees of freedom (df = $N-1$), it is flatter than normal
distribution, hence more prone to producing values far away from the mean (see Fig. 9.8a).

For $N \to \infty$ it becomes the normal distribution. A t-value can be calculated, indicating in terms of standard deviation σ_S, how significantly the sample mean M_S differs from the test mean μ, which is assumed to be the mean value of the source (**one sample t-test**). For N samples m_i:

$$t = \frac{M_S - \mu}{\sigma_S} \sqrt{N} \, \text{with} \, M_S = \frac{1}{N} \sum_{i=1}^{N} (m_i - \mu)^2, \; \sigma_S = \sqrt{\frac{1}{N-1} \sum_{i=1}^{N}}.$$

Usually, it is considered significantly different at the 0.05 level if the sample mean is located in the 2.5% area of the t-distribution on either tail (two-tailed t-test) or significantly greater/less than the test mean if it is in the 5% area in the right/left tail (one-tailed t-test) (see Fig. 9.8b). The **paired-sample t-test** is used for comparing two sets of sample populations, in which the individual subjects are related in pairs (here the $I_{K\text{-}sens}$ and the sum of the two current components $I_K + I_{\text{pump}}$). Hence, the pair-wise difference can be calculated, and then a one sample t-test can be performed on the difference values with the test mean being zero ($\mu = 0$).

9.2.5.3 Determination of the Membrane Capacitance

Procedure
In the window *Export settings* in the *Online* monitor the "Export raw data" option needs to be selected. This will result in an exported file that does not only contain the steady-state IV dependencies, but also the entire time course of current and voltage for each single pulse.

For the determination of membrane capacitance an unloaded oocyte may be used. The cell will be placed in the cell chamber and impaled with the microelectrodes as in the previous type of experiments. Since the membrane capacity is independent of external solution, the chamber can be perfused with a solution of your choice.

Running in *Execution* the *IV-curve_cap* option will apply a single pulse from the holding potential of -60 mV to -70 mV (duration 200 ms) to the membrane.

Tasks
(a) Measure the diameter of the oocyte used for this experiment under the microscope and calculate the total membrane capacitance assuming a ball shape.
(b) Determine the capacitance of the oocyte by analysing the transient current. The file for the single pulse, stored by the CellWorks program consists of three columns: time (ms), current (nA), and potential (mV). At the bottom of the file you find another data set with the steady-state values of potential and current as determined in the previous type of experiment.

Plot the time course of the current. Use Eq. (9.3a) to calculate the membrane capacitance for the potential step from -60 mV to -70 mV and back to -60 mV (take the mean value). What is the corresponding surface area?

(c) Compare and discuss the results of (a) and (b).

References

Bierwirtz, A., & Schwarz, W. (2014). *Biophysikalisches Praktikum*. Goethe Universität. http://www. biophysik.org/~wille/prakt/anleitungen/03_elektrophys.pdf

Fields, R. D. (2007). The Shark's electric sense. *Scientific American, 8*, 75–81.

Guidelines ICNIRP. (1998). Guidelines for limiting exposure to time-varying electric, magnetic and electromagnetic Fields (up to 300 GHz). *Health Physics, 74*, 494–522.

Guidelines ICNIRP. (2010). Guidelines for limiting exposure to time-varying electric and magnetic fields (1 Hz – 100 kHz). *Health Physics, 99*, 818–836.

Guidelines ICNIRP. (2014). Guidelines for limiting exposure to electric fields induced by movement of the human body in a static magnetic field and by time-varying magnetic Fields below 1 Hz. *Health Physics, 106*, 418–425.

Hodgkin, A. L., & Huxley, A. F. (1952). A quantitative description of membrane current and its application to conductance and excitiation in nerve. *The Journal of Physiology, 117*, 500–544.

Jaffe, L. F., & Nuccitelli, R. (1974). An ultrasensitive vibrating probe for measuring steady extracellular currents. *The Journal of Cell Biology, 63*, 614–628.

Lowry, R. (1999). Concepts & applications of inferential statistics. http://vassarstats.net/textbook/

Maeda, K., Henbest, K. B., Cintolesi, F., Kuprov, I., Rodgers, C. T., Liddell, P. A., Gust, D., Timmel, C. R., & Hore, P. J. (2008). Chemical compass model of avian magnetoreception. *Nature, 453*, 387–390.

Mäntele, W. (2021). *Elektrosmog und Ökoboom*. Springer.

Murray, R. W. (1962). The response of the ampullae of Lorenzini of elasmobranchs to electrical stimulation. *Journal of Experimental Biology, 39*, 119–128.

Neher, E., & Sakmann, B. (1976). Single-channel currents recorded from membrane of denervated frog muscle fibres. *Nature, 260*, 799–802.

Robinson, K. R. (1979). Electrical currents through full-grown and maturating Xenopus oocytes. *PNAS, 76*, 837–841.

Schüler, D. (2008). Genetics and cell biology of magnetosome formation in magnetotactic bacteria. *FEMS Microbiology Reviews, 32*, 654–672.

Vistnes, A. I. (2001). Low frequency fields. In D. Brune, R. Hekkborg, B. R. R. Persson, & R. Pääkkönen (Eds.), *Radiation at home, outdoors and in the working place*. Scandinavian Science Publisher.

Correction to: Introduction

Correction to:
Chapter 1 in: J. Rettinger et al.,
Electrophysiology,
https://doi.org/10.1007/978-3-030-86482-8_1

The original version of Chapter 1 was revised. The authors have included four video clips after publication which would add value to the book. The four video files are cited in Chapter 1, at the end of the first paragraph as below:

For illustrating practical application of Two-Electrode Voltage Clamp, we prepared a supplementary video (Videos 1–4) using the *Xenopus* oocytes as a cell model system.

The updated online version of this chapter can be found at
https://doi.org/10.1007/978-3-030-86482-8_1

J. Rettinger et al., *Electrophysiology*,
https://doi.org/10.1007/978-3-030-86482-8_10

Index

Printed in the United States
by Baker & Taylor Publisher Services